Rainfall Erosivity in Soil Erosion Processes

Rainfall Erosivity in Soil Erosion Processes

Special Issue Editors

Gianni Bellocchi

Nazzareno Diodato

MDPI • Basel • Beijing • Wuhan • Barcelona • Belgrade • Manchester • Tokyo • Cluj • Tianjin

Special Issue Editors

Gianni Bellocchi
UCA, INRAE, VetAgro Sup,
UREP
France

Nazzareno Diodato
Met European Research
Observatory
Italy

Editorial Office
MDPI
St. Alban-Anlage 66
4052 Basel, Switzerland

This is a reprint of articles from the Special Issue published online in the open access journal *Water* (ISSN 2073-4441) (available at: https://www.mdpi.com/journal/water/special_issues/soil_erosion_rain).

For citation purposes, cite each article independently as indicated on the article page online and as indicated below:

LastName, A.A.; LastName, B.B.; LastName, C.C. Article Title. *Journal Name* **Year**, *Article Number*, Page Range.

ISBN 978-3-03928-804-5 (Pbk)
ISBN 978-3-03928-805-2 (PDF)

Cover image courtesy of Nazzareno Diodato from the Met European Research Observatory (Benevento, Italy).

Contents

About the Special Issue Editors

Gianni Bellocchi holds a Ph.D. in agronomy and environment from the Sant'Anna School of Advanced Studies (Pisa, Italy). He is currently a senior scientist at INRAE (Institut national de recherche pour l'agriculture, l'alimentation et l'environnement). With expertise in agro-climatic and hydrological modelling, his research and publication interests include carbon–nitrogen cycles and biodiversity in natural and agricultural ecosystems with the support of modelling approaches.

Nazzareno Diodato holds a degree in landscape planning by the University of Naples Federico II. He conducted postgraduate studies at the University Corporation for Atmospheric Research. With focus on the hydrosphere and geosphere, he conducts climate-related studies on environmental monitoring and natural hazards at the Met European Research Observatory (MetEORBS), awarded Laureate in 2011 by the ComputerWorld Honors Program, of which he has been the first administrator since 1986. He has been a committee member of the Commission on Ecosystem Management of the International Union for Conservation of Nature for over 15 years; he is fellow member of the Royal Meteorological Society.

water

MDPI

Editorial

Rainfall Erosivity in Soil Erosion Processes

Gianni Bellocchi [1,*] and Nazzareno Diodato [2]

1 UCA, INRAE, VetAgro Sup, UREP, 63000 Clermont-Ferrand, France
2 Met European Research Observatory, 82100 Benevento, Italy; scodalabdiodato@gmail.com
* Correspondence: gianni.bellocchi@inrae.fr; Tel.: +33-4-4376-1601

Received: 4 March 2020; Accepted: 5 March 2020; Published: 6 March 2020

Abstract: Regional studies on the erosive power of rainfall patterns are still limited and the actual impacts that may follow on erosional and sedimentation processes are poorly understood. Given the several interrelated challenges of environmental management, it is also not always unclear what is relevant for the development of adaptive and integrated approaches facilitating sustainable water resource management. This editorial introduces the Special Issue entitled "Rainfall Erosivity in Soil Erosion Processes", which offers options to fill some of these gaps. Three studies performed in China and Central Asia (by Duulatov et al., *Water* **2019**, *11*, 897, Xu et al., **2019**, *11*, 2429, Gu et al. **2020**, *12*, 200) show that the erosion potential of rainfall is increasing in this region, driving social, economic, and environmental consequences. In the same region (the Weibei Plateau in China), Fu et al. (*Water* **2019**, *11*, 1514) assessed the effect of raindrop energy on the splash distance and particle size distribution of aggregate splash erosion. In the Mediterranean, updated estimates of current and future rainfall erosivity for Greece are provided by Vantas et al. (*Water* **2020**, *12*, 687), while Diodato and Bellocchi (*Water* **2019**, *11*, 2306) reconstructed and investigated seasonal net erosion in an Italian catchment using parsimonious modelling. Then, this Special Issue includes two technologically oriented articles by Ricks at al. The first (*Water* **2019**, *11*, 2386) evaluated a large-scale rainfall simulator design to simulate rainfall with characteristics similar to natural rainfall. The data provided contribute to the information that may be useful for the government's decision making when considering landscape changes caused by variations in the intensity of a rainfall event. The second article (*Water* **2020**, *12*, 515) illustrated a laboratory-scale test of mulching methods to protect against the discharge of sediment-laden stormwater from active construction sites (e.g., highway construction projects).

Keywords: erosion control; mulching; net soil erosion; raindrop energy; rainfall erosivity; runoff; sediment yield

---

1. Introduction

Rainfall erosivity is a major driver of sediment and nutrient losses worldwide, which may leave farmers vulnerable to crop failures and lead to unstable equilibrium states in landscapes [1,2]. The exposure of the Earth's surface to aggressive rainfall is a key factor controlling the water erosion in terrestrial ecosystems [3] and other damaging hydrological events, such as floods and flash floods [4]. The occurrence of hydrological extremes and the associated sediment loss during rainfall events are central features in the global climate system because worldwide variations in temperature and precipitation patterns produce corresponding changes in the development of natural hazards [5]. It is also assumed that extreme storms and rainfall-runoff erosivity are becoming more frequent due to climate change [6]. Highly vulnerable areas may result in catastrophic regime shifts connected with the occurrence of damaging hydrological events [7]. This explains the continuing interest of scientists and engineers in the hydrological response of landscapes. This interest ranges from a basic understanding of processes to prediction under changing conditions, driven by a greater recognition of the cost (both financial and environmental) of neglecting the hydroclimatic forcing factors in relation to soil conservation systems and land-use planning. This Special Issue is an overview of the research and implications for environmental monitoring and policymaking, and encourages further methodological development.

2. Special Issue Overview

The special issue of Water entitled "Rainfall Erosivity in Soil Erosion Processes" publishes eight articles that provide insights into challenges of hydrology and emerging issues at the interface with other related sciences like geomorphology.

The Special Issue contains four articles with a focus on hydrological hazards across China and Central Asia, dealing with: (1) spatiotemporal patterns of rainfall erosivity (Gu et al. [8]), (2) the explanatory power of peak rainfall amounts on sediment yield (Xu et al. [9]), (3) climate-change-induced rainfall erosivity (Duulatov et al. [10]), and (4) rainfall energy-induced soil splash erosion (Fu et al. [11]). The Tibetan Plateau (the focus of Gu et al. [8]), the most active geological belt in China, is not only being affected by the melting of glaciers and other ice formations but also by heavy precipitations that provoke widespread soil loss. These conditions may increase the risk of soil erosion. Especially in spring, rainfall erosivity has been significantly increasing from the 1980s. The southeastern region, where severe soil erosion restricts the development of agriculture and animal husbandry, requires the sustained attention of scientists. The study performed by Xu et al. [9] on the Shixia watershed, in the northeast of the upper reaches of Miyun reservoir in Beijing (China), is about how rainfall morphology affects runoff and soil loss, which is important to deepen the understanding of catchment hydrology and provide support for water and soil resource management. This contribution is worthy because the researches on rainfall affecting runoff and sediment yield rarely analyze impacts from the point of view of rain peak morphology. Fu et al. [11] clarified the effect of raindrop energy on the splash distance and particle size distribution of aggregate splash erosion and introduced a modelling approach to predict splash erosion in the Loess Plateau (central China). Duulatov et al. [10] estimated the potential influence of climate change on erosivity precipitations over Central Asia. In recent years, climatic conditions in Central Asian countries (Kazakhstan, Kyrgyzstan, Tajikistan, Turkmenistan, and Uzbekistan) have changed owing to the reduction in glacier areas, accompanied by a shortage of water for irrigation, degraded natural vegetation covers, erosion processes and salinization, and a decreasing productive capacity of irrigated lands. The authors' predictions indicate that Kyrgyzstan and Tajikistan are expected to be the most affected countries from changes in rainfall patterns, especially from the increase in rainfall erosivity.

Two papers addressed specific topics in the Mediterranean region, where desertification is a serious issue that could be aggravated by rainfall-driven soil erosion. This was highlighted for Greece by Vantas et al. [12], while Diodato and Bellocchi [13] advocated a parsimonious modelling approach for the reconstruction of past erosion data, with an application to a small Mediterranean basin, whose dynamics are analysed in response to climate variability and land-use changes.

Finally, the topics addressed in two studies developed in Alabama (US) by Ricks et al. [14,15] are interesting and challenging in equal measure, and with a high degree of novelty that decision makers may find motivating and engaging. The methods used are high-tech (e.g., adding soil-specific polyacrylamide to erosion-control practices for a greater erosion control), supported by an adequately described mathematical background, and combined with application for construction sites apart from soil erosion itself, the transported sediments being carriers of contaminant factors as well.

3. Conclusions

Water, which is a precious resource for ecosystems, can also turn into a land-disturbing factor due to the erosive force of rainfall, expressed as storm erosivity. Hydrological extremes alter soil structure, triggering erosion, but the ecological consequences of shifts in precipitation extremes and characteristics due to climate change (e.g., nutrient loss and carbon balance) are often poorly understood. The widespread availability of high-temporal resolution rainfall records for large areas and the development of climate models have opened new opportunities for using methods for large scale planning and hazard prevention. This Special Issue raises awareness of the crucial role of hydrological extremes, though the limitations of the body of articles it publishes should be highlighted. It is important that it takes part in sketching the future of two regions particularly sensitive to hydrological changes (i.e., the Mediterranean and Central Asia with China), but there is still a lack of established studies in several regions of the world. Then, this Special Issue does not consider in full measure the extent to which population density, infrastructures, plant density, and other factors influence the occurrence of hydrological damages. The approaches and data resources that this Special Issue introduces are thus expected to promote future research and encourage the consideration of a wide array of scientific sources and possible methods for delivering decision support in various contexts.

Author Contributions: G.B. and N.D. wrote this editorial for the introduction of the Special Issue, entitled "Rainfall Erosivity in Soil Erosion Processes", of Water, and edited the Special Issue. All authors have read and agreed to the published version of the manuscript.

Funding: This research received no external funding.

Acknowledgments: We thank all authors who submitted their valuable papers to the Special Issue, entitled "Rainfall Erosivity in Soil Erosion Processes", of Water. The editors also acknowledge the dedication and diligence of the qualified reviewers who provided valuable and insightful comments that helped to direct the articles featured in this special issue and contributed to improving the content.

Conflicts of Interest: The authors declare no conflict of interest.

References

1. Panagos, P.; Borrelli, P.; Meusburger, K.; Bofu, Y.; Klik, A.; Jae Lim, K.; Yang, J.E.; Ni, J.; Miao, C.; Chattopadhyay, N.; et al. Global rainfall erosivity assessment based on high-temporal resolution rainfall records. *Sci. Rep.* **2017**, *7*, 4175. [CrossRef] [PubMed]
2. Wuepper, D.; Borrelli, P.; Finger, R. Countries and the global rate of soil erosion. *Nature Sustain.* **2020**, *3*, 51–55. [CrossRef]
3. Li, Z.Y.; Fang, H.Y. Impacts of climate change on water erosion: A review. *Earth-Sci. Rev.* **2016**, *163*, 94–117. [CrossRef]
4. Diodato, N.; Borrelli, P.; Fiener, P.; Bellocchi, G.; Romano, N. Discovering historical rainfall erosivity with a parsimonious approach: A case study in Western Germany. *J. Hydrol.* **2017**, *544*, 1–9. [CrossRef]
5. Brönnimann, S.; Pfister, C.; White, S. *The Palgrave Handbook of Climate History*; White, S., Pfister, C., Mauelshagen, F., Eds.; Palgrave Macmillan: Basingstoke, UK, 2018; pp. 27–36.
6. Yin, J.; Gentine, P.; Zhou, S.; Sullivan, S.C.; Wang, R.; Zhang, Y.; Guo, S. Large increase in global storm runoff extremes driven by climate and anthropogenic changes. *Nat. Commun.* **2018**, *9*, 4389. [CrossRef] [PubMed]
7. Diodato, N.; Borrelli, P.; Panagos, P.; Bellocchi, G.; Bertolin, C. Communicating hydrological hazard-prone areas in Italy with geospatial probability maps. *Front. Environ. Sci.* **2019**, *7*, 193. [CrossRef]
8. Gu, Z.; Feng, D.; Duan, X.; Gong, K.; Li, Y.; Yue, T. Spatial and temporal patterns of rainfall erosivity in the Tibetan Plateau. *Water* **2020**, *12*, 200. [CrossRef]
9. Xu, J.; Zhang, J.; Li, M.; Wang, F. Effect of rain peak morphology on runoff and sediment yield in Miyun water source reserve in China. *Water* **2019**, *11*, 2429. [CrossRef]
10. Duulatov, E.; Chen, X.; Amanambu, A.C.; Ochege, F.U.; Orozbaev, R.; Issanova, G.; Omurakunova, G. Projected rainfall erosivity over Central Asia based on CMIP5 climate models. *Water* **2019**, *11*, 897. [CrossRef]
11. Fu, Y.; Li, G.; Wang, D.; Zheng, T.; Yang, M. Raindrop energy impact on the distribution characteristics of splash aggregates of cultivated dark loessial cores. *Water* **2019**, *11*, 1514. [CrossRef]
12. Vantas, K.; Sidiropoulos, E.; Loukas, A. Estimating current and future rainfall erosivity in Greece using regional climate models and spatial quantile regression forests. *Water* **2020**, *12*, 687. [CrossRef]
13. Diodato, N.; Bellocchi, G. Reconstruction of seasonal net erosion in a Mediterranean landscape (Alento River Basin, Southern Italy) over the past five decades. *Water* **2019**, *11*, 2306. [CrossRef]

14. Ricks, M.D.; Horne, M.A.; Faulkner, B.; Zech, W.C.; Fang, X.; Donald, W.N.; Perez, M.A. Design of a pressurized rainfall simulator for evaluating performance of erosion control practices. *Water* **2019**, *11*, 2396. [CrossRef]
15. Ricks, M.D.; Wilson, W.T.; Zech, W.C.; Fang, X.; Donald, W.N. Evaluation of hydromulches as an erosion control measure using laboratory-scale experiments. *Water* **2020**, *12*, 515. [CrossRef]

water

MDPI

Article

Estimating Current and Future Rainfall Erosivity in Greece Using Regional Climate Models and Spatial Quantile Regression Forests

Konstantinos Vantas *, Epaminondas Sidiropoulos and Athanasios Loukas

Department of Rural and Surveying Engineering, Aristotle University of Thessaloniki, 54124 Thessaloniki, Greece; nontas@topo.auth.gr (E.S.); agloukas@topo.auth.gr (A.L.)

* Correspondence: kon.vantas@gmail.com or vantas@topo.auth.gr; Tel.: +30-24670-24804

Received: 14 February 2020; Accepted: 1 March 2020; Published: 3 March 2020

check for updates

Abstract: A future variation of precipitation characteristics, due to climate change, will affect the ability of rainfall to precipitate soil loss. In this paper, the monthly and annual values of rainfall erosivity (R) in Greece are calculated, for the historical period 1971–2000, using precipitation records that suffer from a significant volume of missing values. In order to overcome the data limitations, an intermediate step is applied using the calculation of monthly erosivity density, which is more robust to the presence of missing values. Spatial Quantile Regression Forests, a data driven algorithm that imitates kriging without the need of strict statistical assumptions, was utilized and validated, in order to create maps of R and its uncertainty using error propagation. The monthly average precipitation for the historical period 1971–2000 estimated by five (5) Global Circulation Models-Regional Climatic Models were validated against observed values and the one with the best performance was used to estimate projected changes of R in Greece for the future time period 2011–2100 and two different greenhouse gases concentration scenarios. The main findings of this study are: (a) the mean annual R in Greece is 1039 MJ·mm/ha/h/y, with a range between 405.1 and 3160.2 MJ·mm/ha/h/y. The highest values are calculated at the mountain range of Pindos and the lowest at central Greece; (b) the monthly R maps adhere to the spatiotemporal characteristics of precipitation depth and intensities over the country; (c) the projected R values, as an average over Greece, follow the projected changes of precipitation of climatic models, but not in a spatially homogenous way.

Keywords: rainfall erosivity; erosivity density; climate change; regional climate models; quantile regression forests; Greece

1. Introduction

Rainfall erosivity concerns the ability of rainfall to precipitate soil loss [1], as it supplies energy to the mechanical processes of soil erosion. Decertification has been identified as one of the most serious issues facing Mediterranean European countries, including Greece [2], and a possible increase in future rainfall, due to climate change, will aggravate this process, as soil erosion increases at a greater rate [3]. Unanticipatedly, a decrease in future rainfall and a possible decrease of biomass production may also lead to higher erosion rates [4].

Higher erosion rates in conjunction with unsustainable land management and increasing human pressure can lead to soil degradation [5], and consequently a disrupted ecological balance, a decreasing agricultural production and income [6] and even the reduction of effectiveness of adaptation options [7]. Several issues may arise due to accelerated soil losses on achieving of the Sustainable Development Goals of the United Nations [8], as these goals are dependent on a healthy biophysical environment in which the soil is the base [9]. In order to predict these soil erosion future changes it is necessary to simulate changes in future rainfall erosivity, land uses and the application of policies on land management [10].

Universal Soil Loss Equation (USLE) [11], which is the most widely used soil erosion prediction model in the world [12], is an empirical equation that estimates the long-term, average, rate of soil loss involving the product of six factors. The USLE erosivity factor, *R*, is calculated using high frequency or break-point precipitation data with a duration of over 20 years [13,14], as a function of rainfall intensity and depth. In the second revised version of USLE, RUSLE2 [15], monthly Erosivity Density (*ED*) was introduced, as a measure of rainfall erosivity per unit rainfall, which requires shorter precipitation record lengths. *ED* is approximately a function of values only related to rainfall intensity and was used in RUSLE2 as an intermediate step, in conjunction with coarser, monthly, precipitation data, to compute R values in the USA.

Precipitation in Greece has been investigated in several studies over the past two decades [16–23]. In general, precipitation varies from its maximum values during winter to a minimum during summer. The highest precipitation values are observed on the mountain range of Pindos, and its expansion on Peloponnesus, and the lowest values of precipitation are recorded on the Cyclades Islands, at the center of the Aegean Sea. Due to the fact that most weather systems and prevailing winds are moving over the Ionian Sea perpendicularly to Pindos, a contrast exists between the wetter western parts of the country and the dryer eastern ones. During summer months convective activity over northern Greece produces higher precipitation amounts than over the drier southern parts.

Global Circulation Models (GCMs) are models that represent the atmosphere, land surface, ocean and sea ice and simulate their interactions in three dimensions, to make long-term predictions of climate [24]. Different scenarios about future concentrations of greenhouse gases (Representative Concentration Pathways, RCPs) are employed to describe a set of different climate futures that drive GCMs [25]. Due to the coarse grid scale of GCMs (over 80 to ~300 km), Regional Climate Models (RCMs) were developed to downscale them and provide information on a finer scale, more applicable to local scaled phenomena, impact studies and adaptation decisions. RCMs are dependent on GCMs, because GCMs provide the response of global circulation, greenhouse gases concentrations, etc. and RCMs refine them in a spatiotemporal sense, using features such as the topography, coastlines, land cover or mesoscale dynamics [26]. The climatic models from COordinated Regional Climate Downscaling EXperiment over Europe (EURO-CORDEX) [27], using high-resolution RCMs for the high greenhouse gases concentration scenario, RCP8.5, project a decrease of precipitation from 1971–2000 to 2071–2100 and for the medium concentration scenario, RCP4.5, project the same trend with a smaller magnitude [27,28]. Regarding rainfall intensity, in the form of heavy precipitation that exceeds the intensity at the 95th percentile of daily precipitation, the same models project diverging trends that are not statistically significant in most areas of Greece [27,28].

In our previous studies about *ED* in Greece [29,30] it was proven that, in general, *ED* values are robust to the presence of missing values in contrast to *R*, which is highly affected, and specifically in Greece: (a) the values of *ED* are not significantly correlated with the elevation, (b) *ED* annual timeseries are found to be stationary, in contrast to reported precipitation trends for the same time period and (c) *ED* can be considered as spatially autocorrelated, as three contiguous areas were identified using clustering analysis, that had distinct temporal patterns. A comparison of our studies with an earlier study about R in Greece by Panagos et al. [31] revealed that the previously reported R values were underestimated due to the presence of a significant volume of missing data in the precipitation records used in the calculations.

In the Mediterranean region, the annual *R* model MedREM was developed using annual precipitation depth, the longitude and annual daily maximum precipitation data [32]. A recent paper regarding the estimation of future *R* values in Europe for 2050 [10], used one GCM and a single RCP, applied Gaussian Process Regression using monthly variables obtained from the WorldClim dataset [33] and estimated an increase of *R* by 14.8% in Greece. A number of papers in Europe examined the potential increase of rainfall erosivity using temporal trends of high resolution precipitation data in Western Germany [34], Belgium [35] and in the Czech Republic [36]. Other studies in various parts of the world used GCMs in conjunction with empirical equations that predict R using annual precipitation [37,38], monthly [39,40] and daily rainfall indices [41,42]. A different approach estimated projected *R* changes, using a weather generator with spatial and temporal downscaled precipitation values coming from various GCMs [43].

Random Forests [44] is a data-driven algorithm in the area of supervised learning which tries to fit a model using a set of paired input variables and their associated output response and can be used in classification and regression problems. Quantile Regression Forests (QRF) [45], is an extension of RF that is able to compute prediction intervals of the output response in regression problems. RF has been used for spatial prediction in various domains [46–50] and recently, Hengl et al. [51] presented RF for a spatial predictions framework, that can make equally accurate predictions as kriging, without the need of statistical assumptions.

The aim of this work is to calculate the current and estimate the future changes of *R* values in Greece. The latest methodologies developed and presented with RUSLE2 are used, taking into account the presence of missing values in precipitation records. The first objective of the analysis is to create monthly precipitation and *ED* maps, as intermediate datasets, and to estimate the uncertainty of their predictions and their errors using cross-validation. Consequently, the current *R* values in Greece are computed from the interpolated surfaces and error propagation is used to estimate approximately their uncertainty. Finally, downscaled precipitation from GCMs-RCMs is validated and used, along with *ED*, to estimate the potential changes of R in Greece for the years 2040, 2070 and 2100 using two future greenhouse gases concentration trajectories/pathways (i.e., RCPs). This type of analysis is a novel approach and it has not been presented, until now, in the international literature.

2. Materials and Methods

The methodology that was applied in the study, is presented in the flowchart of Figure 1, where data flows from the left to the right. Blue symbolizes the data that were used as input (pointwise and raster), orange the intermediate raster datasets that were created and green the final results, also in raster format, that were computed using different sets of input and intermediate datasets.

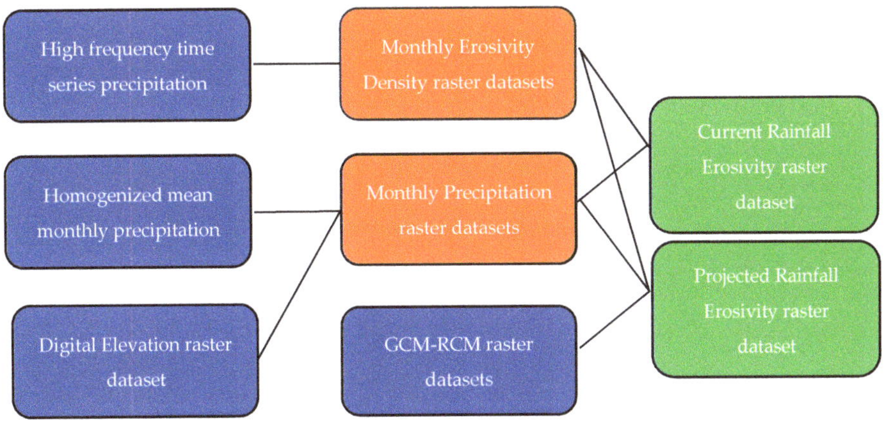

Figure 1. Flowchart of the applied methodology.

2.1. Data Acquisition and Processing

Point precipitation data from 237 meteorological stations across Greece (Figure 2), was used in the analysis. The data consisted of:

- Pluviograph data for 87 meteorological stations that were taken from the Greek National Bank of Hydrological and Meteorological Information (NBHM) [52]. The time series comprised a total of 2273 years of 30-min-records and 394 years of five-min-records for the time period from 1953 to 1997, with a mean length of 30 years per station. The timeseries coverage was 62.8% on average.
- Mean monthly precipitation data for 150 meteorological stations that were taken from the Hellenic National Meteorological Service (HNMS) [53]. These are homogenized data and available for the time period from 1971 to 2000. These data were used to overcome the limitations of precipitation data from NBHM.
- Five different monthly GCM-RCM raster datasets were downloaded for a number of experiments and time periods (Table 1). These data were computed in the framework of the EURO-CORDEX [27,54], had a horizontal resolution 0.11° × 0.11° and were remapped using bilinear interpolation to a 30″ × 30″ resolution grid using the Climate Data Operator (CDO) software [55].

Table 1. Global Circulation Models (GCMs)-Regional Climate Models (RCMs) used in the analysis (i.e., data retrieved from EURO-CORDEX).

	GCM	RCM	Institution
1	EC-EARTH	DMI-HIRHAM5	Danish Meteorological Institute
2	EC-EARTH	KNMI-RACMO22E	Royal Netherlands Meteorological Institute
3	HadGEM2-ES	KNMI-RACMO22E	Royal Netherlands Meteorological Institute
4	MPI-ESM-LR	CSC-REMO2009	Max Planck Institute for Meteorology
5	NorESM1-M	DMI-HIRHAM5	Danish Meteorological Institute

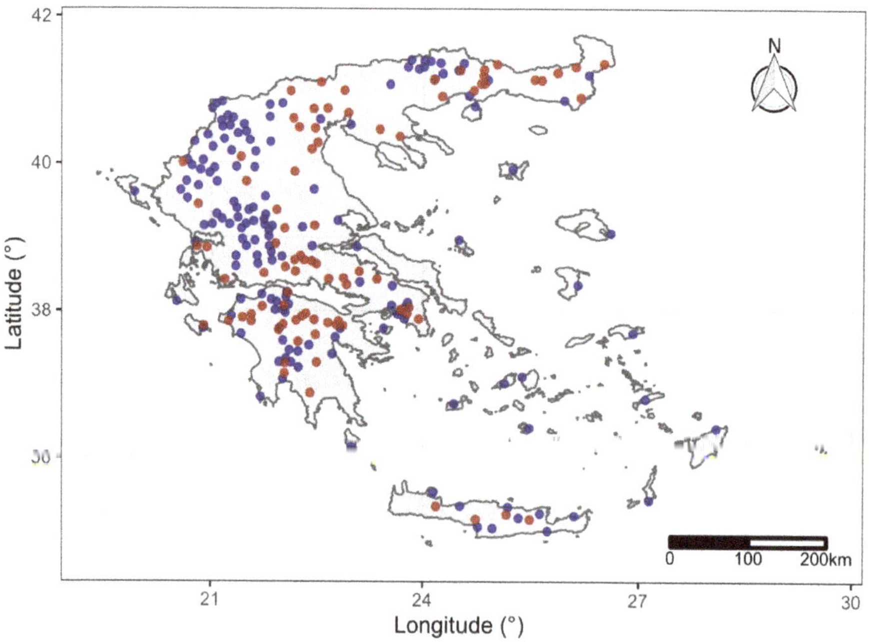

Figure 2. Station locations in Greece used in the analysis. Red points symbolize the 87 stations with pluviograph data from the Greek National Bank of Hydrological and Meteorological Information (NBHM) and blue points are the 150 stations with average monthly precipitation from the Hellenic National Meteorological Service (HNMS).

The GCM-RCM monthly precipitation timeseries data were:

- Historical, for the time period from 1971 to 2000 (like the ones coming from HNMS) as they were driven by the boundary conditions provided by the GCMs.
- Future, for the forcing scenarios RCP4.5 (where greenhouse gases emissions peak around 2040 and then decline) and RCP8.5 (where greenhouse gases emissions continue to rise throughout the 21st century), using the control ensemble of the CORDEX climate projection experiments. The dataset period was from 2011 to 2100.

Digital elevation raster data were downloaded from the NASA Shuttle Radar Topography Mission (SRTM) [56], and aggregated to the same 30″ × 30″ resolution grid as the one used in the GCM-RCMs datasets.

2.2. Monthly Erosivity Density Calculation

The erosivity of a single erosive rainfall event, EI_{30} (MJ·mm/ha/h), given the product of the kinetic energy of rainfall and its maximum 30 min intensity, was computed using the pluviograph records from NBHM [15]:

$$EI_{30} = \left(\sum_{r=1}^{m} e_r \cdot v_r\right) \times I_{30} \tag{1}$$

where e_r is the kinetic energy per unit of rainfall (MJ/ha/mm), v_r the rainfall depth (mm) for the time interval r of the hyetograph, which has been divided into $r = 1, 2, \ldots, s$ time sub-intervals and I_{30} is the maximum rainfall intensity for a 30 min duration during that rainfall.

The quantity e_r was calculated for each time sub-interval, r, applying the kinetic energy equation that was used in RUSLE2 [57], which was recently evaluated in Italy, a nearby country to Greece, and had the best performance among alternative literature expressions [58]:

$$e_r = 0.29 \cdot \left(1 - 0.72e^{-0.82i_r}\right) \tag{2}$$

where i_r is the rainfall intensity (mm/h).

An individual rainfall event was extracted from the continuous pluviograph data, if its cumulative depth for a duration of 6 h at a certain location was less than 1.27 mm. A rainfall event was considered to be erosive if it had a cumulative rainfall depth greater than 12.7 mm. Only the screened events with a return period of less than 50 years were used in the calculations.

On the grounds that the use of coarser fixed time intervals to a finer one can lead to an underestimation of the value of erosivity [59,60], monthly conversion factors c_m were computed using the five-min-time-step timeseries:

$$c_m = \frac{1}{n_m} \sum_{i=1}^{n_m} \frac{(EI_{30})_{m,\ ts=5\ \text{min}}}{(EI_{30})_{m,\ ts=30\ \text{min}}} \tag{3}$$

where n_m is the number of storms at month m, $(EI_{30})_{m,\ ts=5\ \text{min}}$ is the erosivity of a storm using the five min time step and $(EI_{30})_{m,\ ts=30\ \text{min}}$ the erosivity of the same storm when the timeseries was aggregated using a 30 min time step. These conversion factors were applied to the values of erosivity that were estimated from 30-min pluviograph data.

After the computation of EI_{30} values, the average monthly rainfall erosivity density ED_m (MJ/ha/h) per station was calculated:

$$ED_m = \frac{1}{n} \sum_{i=1}^{n} \left(\frac{\sum_{k=1}^{st_m} (EI_{30})_k}{P_m}\right)_i \tag{4}$$

where st_m is the number of storms during the month m, $(EI_{30})_k$ the erosivity of storm k, P_m the monthly precipitation height and n the number of years.

2.3. Spatial Quantile Regression Forests

Random Forests (RF) [44] is one of the most successful methods used in Machine Learning [61], among other reasons because of: (a) its robustness to outliers [62] and overfitting [63], (b) its ability to perform feature selection [64] and (c) the fact that its default parameters, as implemented in software, give satisfactory results [61,65]. Examples of RF open source software are the R's language packages randomForest [66] and its faster alternative, ranger [67].

In summary, RF consists of a number of decision trees [68]. For each tree, a random set of the dataset is created via bootstrapping [69] and in each node of the tree a random set of n input variables from the p variables of the dataset is considered to pick the best split [70]. The prediction of the output response in regression problems is the mean value of the estimations of these random decision trees. The estimate of the out-of-sample error is computed using the out-of-bag error [71], without the need of cross-validation.

Quantile Regression Forests (QRF) is an extension of RF that provides information about the full conditional distribution of the output response and not only about its mean [45], as is the case in plain RF. In this way, it is possible to provide prediction intervals and measures of uncertainty.

The use of QRF as a framework for the modeling of spatial variables was introduced by Hengl et al. [51], where the distances among observation locations are used as variables in QRF in order to incorporate geographical proximity effects. In this way, using these buffer distances, spatial QRF imitate kriging's spatial correlation. Spatial QRF (spQRF) produces comparable results, in terms of accuracy, compared to the state-of-the-art kriging methods, with the advantage of no prior assumptions about the distribution or stationarity of the response variable [51].

In this work, spQRF were used for the spatial prediction of monthly precipitation and *ED*:

- The 12 monthly *ED* models (one model for each month of the year) were trained using as input variables the buffer distances for each one of the 87 stations of NBHM in a chained procedure [72]. In that procedure, at the first run, only the distances are used as input variables and at the nth both distances and the monthly results from the previous step, with the exception of the month that is the output response. This procedure stops when the out-of-sample error estimated by RFs in the form of out-of-the-bag error ceases to decrease.
- The 12 monthly precipitation models were trained using as input variables the buffer distances for each one of the 150 stations of HNMS and elevation data from the SRTM.

As a measure of the out-of-sample error, the average Root Mean Squared Error (*RMSE*), the coefficient of determination R^2 and Lin's Concordance Correlation Coefficient (CCC) were computed using 10-fold cross validation:

$$RMSE = \sqrt{\frac{1}{n}\sum_{i=1}^{n}\left(\hat{y}^{(s_i)} - y^{(s_i)}\right)^2} \tag{5}$$

$$R^2 = 1 - \frac{\sum_{i=1}^{n}\left(\hat{y}^{(s_i)} - y^{(s_i)}\right)^2}{\sum_{i=1}^{n}\left(\hat{y}^{(s_i)} - \mu_y\right)^2} \tag{6}$$

$$CCC = \frac{2 \cdot \sigma_{\hat{y}y}}{\sigma_{\hat{y}}^2 + \sigma_y^2 + \left(\mu_{\hat{y}} - \mu_y\right)^2} \tag{7}$$

where n is the total number of cross-validation locations, $\hat{y}^{(s_i)}$ is the predicted value of $y^{(s_i)}$ at a cross-validation location s_i (i.e., the coordinates longitude and latitude of location i), $\mu_{\hat{y}}$, μ_y, $\sigma_{\hat{y}}^2$, σ_y^2 are the means and variances of $\hat{y}^{(s_i)}$ and $y^{(s_i)}$, respectively, and $\sigma_{\hat{y}y}$ is the covariance of $\hat{y}^{(s_i)}$ and $y^{(s_i)}$.

R^2 describes the ratio of variance that is explained by a model and may be negative, among other reasons, if an inappropriate model is used [73]. *CCC* combines measures of both precision and accuracy and examines how far $\hat{y}$ deviate from the line of perfect concordance (the line of 45 degrees on a square scatterplot) and ranges from 0 to ±1 [74,75].

The uncertainty of the predictions from the models, in the form of prediction error standard deviation, was computed using a dense level of all the quantiles per 1% of the output response and for every location [51]:

$$\sigma^{(S_i)} = \sqrt{\frac{\sum_{p=1\%}^{100\%}\left(\hat{y}_p{}^{(s_i)} - \mu_{\hat{y}^{(s_i)}}\right)^2}{100-1}} \quad (8)$$

where $\hat{y}_p{}^{(s_i)}$ is the pth percentile of the distribution of the response variable at location s_i and $\mu_{\hat{y}^{(s_i)}}$ the mean value of $\hat{y}_p{}^{(s_i)}$ for all the percentiles.

The quantity z-score, which quantifies the error of prediction errors, was calculated at cross validation locations [76]:

$$z^{(s_i)} = \frac{\hat{y}^{(s_i)} - y^{(s_i)}}{\sigma^{(s_i)}} \quad (9)$$

where z, ideally, should have a mean equal to zero and variance equal to one. On the contrary:

- If variance(z) >> 1, the model underestimates the actual prediction uncertainty.
- If variance(z) << 1, the model overestimates the actual prediction uncertainty.

2.4. Regional Climate Models Historical Precipitation Validation

In order to validate and select one of the GCMs-RCMs for the projected erosivity calculations, at first, a multi-layer raster dataset was computed for each of the five models with the overall 30-year-mean-monthly precipitation values, using the historical time period from 1971 to 2000. Then, using the monthly precipitation spQRF models to create raster datasets with the same 30″ × 30″ resolution grid, *RMSE*, *CCC* and R^2 errors metrics were computed, setting in Equations (5)–(7) as $\hat{y}^{(s_i)}$ the values estimated by GCMs-RCMs and as $y^{(s_i)}$ the values calculated by the spQRF models.

2.5. Current and Projected Erosivity Calculation

The estimation of current and future monthly erosivity was made under the assumption of future temporal stationarity of *ED* values, due to the fact that *ED* is related to seasonal rainfall intensity [15] and EURO-CORDEX GCMs-RCPs models does not project statistically significant trends in most areas of Greece [27]. Temporal stationarity of *ED* values already has been documented in Greece for the historical period [30].

The current monthly erosivity was calculated as the product of predictions of the trained spQRF models of monthly ED_m and precipitation P_m for each month m:

$$\hat{R}_m{}^{(s_i)} = \hat{P}_m{}^{(s_i)} \cdot \hat{ED}_m{}^{(s_i)} \quad (10)$$

The prediction error standard deviation of monthly R_m was calculated using error propagation [77]:

$$\sigma_{R_m}^{(s_i)} = \hat{R}_m{}^{(s_i)} \cdot \sqrt{\left(\frac{\sigma_{P_m}^{(s_i)}}{\hat{P}_m{}^{(s_i)}}\right)^2 + \left(\frac{\sigma_{ED_m}^{(s_i)}}{\hat{ED}_m{}^{(s_i)}}\right)^2} \quad (11)$$

In order for Equation (11) to hold true, $\hat{P}_m{}^{(s_i)}$ and $\hat{ED}_m{}^{(s_i)}$ were considered as the true values on locations s_i and that were independent of each other. More specifically, these values were considered, to a good approximation, as not correlated on the basis of the following remarks:

- $\hat{P}_m{}^{(s_i)}$ came from monthly average data and $\hat{ED}_m{}^{(s_i)}$ from random proportions of pluviograph data due to missing values.

- In general, precipitation depth, alone, is a poor indicator of erosivity [78].
- Using the approximation that kinetic energy in Equation (1) can be replaced with a constant monthly value [79], then as is proved in RUSLE2 [15], monthly *ED* is directly proportional to 30-min rainfall intensity without the effect of precipitation depth.

The annual rainfall erosivity and its error standard deviation were calculated, respectively:

$$\hat{R}^{(s_i)} = \sum_{m=1}^{12} \hat{R}_m{}^{(s_i)} \tag{12}$$

$$\sigma_R^{(s_i)} = \sqrt{\sum_{m=1}^{12} \left(\sigma_{R_m}^{(s_i)}\right)^2} \tag{13}$$

The projected values of monthly erosivity were computed using the ratio $\delta_{f,m}$ of the future 30-years-mean $P^{(s_i)}_{GCM-RCM,\ future,\ m}$ to the historical 30-years-average-monthly precipitation values $P^{(s_i)}_{GCM-RCM,\ historical,\ m}$ that came from a GCM-RCM:

$$\hat{R}^{(s_i)}{}_{f,\ m} = \left(\delta^{(s_i)}_{f,m} \cdot \hat{P}^{(s_i)}{}_m\right) \cdot \hat{ED}_m{}^{(s_i)} = \delta^{(s_i)}_{f,m} \cdot \hat{R}_m{}^{(s_i)} \tag{14}$$

where f is the future year in which long term average monthly values refer to and:

$$\delta^{(s_i)}_{f,m} = \frac{P^{(s_i)}_{GCM-RCM,\ future,\ m}}{P^{(s_i)}_{GCM-RCM,\ historical,\ m}} \tag{15}$$

The projected annual erosivity was estimated with:

$$\hat{R}^{(s_i)}{}_f = \sum_{m=1}^{12} \delta^{(s_i)}_{f,m} \cdot \hat{R}_m{}^{(s_i)} \tag{16}$$

And the ratio of future to current values:

$$\hat{r}^{(s_i)}{}_f = \frac{\hat{R}^{(s_i)}{}_f}{\hat{R}^{(s_i)}} \tag{17}$$

With Equations (14)–(17), the projected annual erosivity values were calculated preserving the relative projected changes of precipitation without the direct use of simulated values coming from a GCM-RCM and the need to apply a bias-correction method. However, the need to use a statistical downscaling technique to apply bias correction to the GCM + RCM output will be tested.

3. Results and Discussion

3.1. Precipitation and ED Pointwise Values

The monthly conversion factors c_m that were computed specifically for Greece using the five-min-time-step timeseries have a mean value 1.22, close to the one calculated for Europe [59], but had different seasonality with their maximum values in the period from May to October and minimum during March (Table 2).

Table 2. Monthly erosivity conversion factors c_m (unitless) for 30 min timestep compared to the base timestep of five min.

	Jan	Feb	Mar	Apr	May	Jun	Jul	Aug	Sep	Oct	Nov	Dec	Mean
c_m	1.20	1.18	1.16	1.18	1.25	1.24	1.29	1.26	1.24	1.25	1.19	1.19	1.22

The first four central moments (mean, standard deviation, skew and kurtosis) and other statistical properties were used to describe the calculated monthly values of *ED* (87 stations from NBHM, Table 3) and the average monthly precipitation (150 stations from HNMS, Table 4).

Table 3. The average statistical properties of calculated monthly Erosivity Density (*ED*) (MJ/ha/h). SD is an abbreviation for standard deviation and CV for coefficient of variation (the ratio of the standard deviation to the mean).

Prec (mm)	Min	Mean	Median	Max	SD	Skew	Kurtosis	CV
January	0.380	1.138	1.110	2.345	0.435	0.416	−0.455	0.383
February	0.535	1.149	1.088	2.403	0.409	0.787	0.130	0.356
March	0.525	1.113	1.054	2.413	0.356	1.105	1.681	0.320
April	0.464	1.099	1.062	2.175	0.316	0.895	0.876	0.288
May	0.407	1.496	1.404	2.645	0.447	0.377	−0.435	0.299
June	0.850	1.854	1.712	4.137	0.641	0.992	0.953	0.346
July	1.215	2.341	2.102	5.445	0.863	1.381	1.800	0.369
August	0.703	2.079	1.987	5.993	0.819	1.637	4.921	0.394
September	0.912	1.842	1.657	3.786	0.669	1.017	0.427	0.363
October	0.666	1.916	1.791	3.891	0.706	1.018	0.650	0.369
November	0.589	1.732	1.619	3.904	0.677	0.574	−0.103	0.391
December	0.517	1.442	1.435	3.497	0.568	0.680	0.635	0.394

Table 4. The average statistical properties of observed monthly precipitation values (mm). SD is an abbreviation for standard deviation and CV for coefficient of variation (the ratio of the standard deviation to the mean).

Prec (mm)	Min	Mean	Median	Max	SD	Skew	Kurtosis	CV
January	25.8	101.5	99.6	227.8	46.2	0.515	−0.378	0.455
February	34.6	104.9	93.4	257.4	50.5	0.766	−0.096	0.482
March	29.5	83.8	75.0	202.4	35.8	0.802	0.045	0.427
April	18.5	69.9	61.6	183.8	36.8	0.644	−0.371	0.526
May	7.4	51.3	51.6	149.1	29.3	0.342	−0.455	0.571
June	1.0	26.9	26.1	81.9	19.1	0.692	0.097	0.711
July	0.1	23.6	22.8	80.6	18.0	0.877	0.726	0.762
August	0.1	24.5	24.4	78.3	16.7	0.498	0.012	0.679
September	4.4	37.0	36.7	96.9	20.2	0.375	−0.414	0.546
October	30.6	85.9	75.7	205.7	40.6	0.738	−0.172	0.473
November	47.5	134.8	110.4	300.6	66.0	0.556	−0.808	0.490
December	43.9	133.0	117.6	329.2	65.6	0.742	−0.249	0.493

Monthly precipitation and *ED* had different monthly temporal patterns (Figure 3). The temporal pattern of precipitation is typical for a Mediterranean area with the minimum precipitation occurring during summer and the maximum precipitation observed in November and December (Figure 3a). This temporal pattern of monthly precipitation is typical for Greece [80]. The second had an almost bimodal shape with its two peaks at July and October. The values of *ED* were slightly different from the ones already reported in our previous work [30], due to the application of the seasonal monthly conversion factors c_m to a constant one.

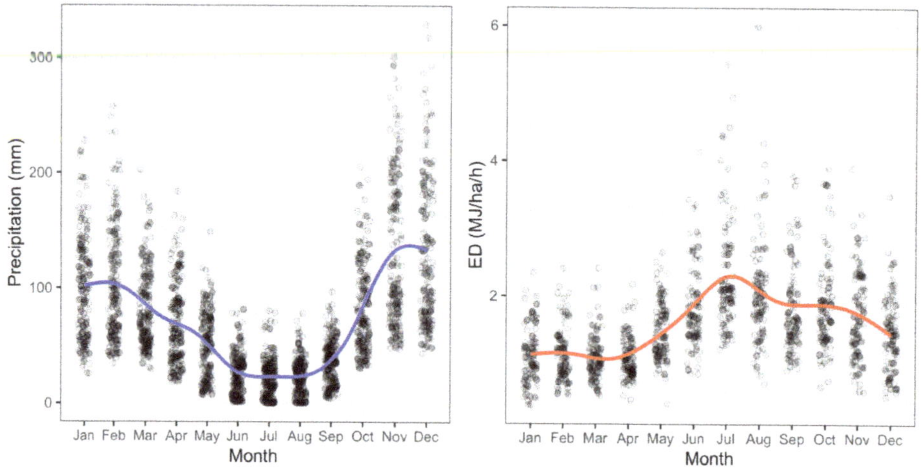

(a) (b)

Figure 3. Scatterplots for: (**a**) Observed mean monthly precipitation values; (**b**) Calculated *ED* monthly values. In the above plots jitter exists in order to make visible all points. With color are the smooth lines produced by means of Local Polynomial Regression Fitting [81].

3.2. Spatial Models of Precipitation and ED

The training and cross validation of the spatial models was made using the implementation of QRF in the ranger [67] package of language *R* [82]. The number of trees was set to 1000 and the fine-tuning of the parameters of the model was made using the tuneRanger [83] package.

The cross validation error metrics of the models for monthly precipitation were satisfactory (Table 5), especially comparing them to recent precipitation models for Greece that showed weak correlation during the spring season [17] or equally good estimations [23]. As most months had a z-score variance smaller than one, these monthly models overestimated the actual prediction uncertainty (i.e., the error of prediction the error from the models in cross-validation locations).

Table 5. Cross validation metrics for monthly precipitation spQRF models. *RMSE* units are in mm, R^2, z-score related values and Concordance Correlation Coefficient (*CCC*) are unitless.

	Jan	Feb	Mar	Apr	May	Jun	Jul	Aug	Sep	Oct	Nov	Dec	Mean
RMSE	24.4	25.9	19.9	15.9	11.8	7.4	6.8	7.6	7.9	16.2	26.6	30.6	16.7
R^2	0.68	0.68	0.62	0.77	0.81	0.83	0.84	0.76	0.83	0.81	0.81	0.75	0.77
CCC	0.81	0.83	0.79	0.87	0.89	0.90	0.91	0.86	0.90	0.90	0.90	0.86	0.87
$\bar{z}$	−0.11	−0.08	−0.03	−0.03	0.01	0.03	−0.02	−0.07	−0.04	−0.04	−0.07	−0.07	−0.05
σ_z	1.01	0.87	0.89	0.73	0.69	0.64	0.58	0.68	0.77	0.92	1.10	0.87	0.81

The same cross validation error metrics of the models for monthly *ED* did not show equally good performance (Table 6), as most months had moderate results with the exception of summer months that had poor results. These poor results were coming from the scarcity of the stations used and the predictions of high *ED* values that the models underestimated them (Figure 4b), affecting especially *RMSE* and R^2 metrics that are sensitive to outliers. Given the values of z-score's variance, in general, also, most of the monthly *ED* models overestimated the actual prediction uncertainty.

Table 6. Cross validation metrics for monthly *ED* Spatial Quantile Regression Forests (spQRF) models. *RMSE* units are in MJ/ha/h, R^2 z-score related values and *CCC* are unitless.

	Jan	Feb	Mar	Apr	May	Jun	Jul	Aug	Sep	Oct	Nov	Dec	Mean
RMSE	0.29	0.29	0.27	0.22	0.35	0.48	0.71	0.67	0.46	0.44	0.39	0.33	0.41
R^2	0.38	0.37	0.23	0.37	0.02	0.07	0.06	0.13	0.38	0.49	0.46	0.36	0.28
CCC	0.62	0.56	0.51	0.55	0.51	0.52	0.39	0.38	0.59	0.66	0.74	0.73	0.56
$\bar{z}$	−0.07	−0.15	−0.08	−0.10	−0.06	−0.10	−0.10	−0.10	−0.06	−0.07	−0.06	−0.08	−0.09
σ_z	1.09	1.00	0.79	0.95	0.70	1.21	1.10	1.01	0.82	0.79	0.76	0.68	0.91

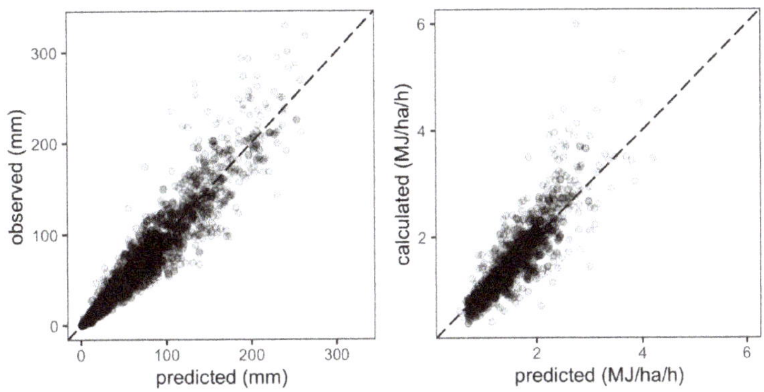

(a) (b)

Figure 4. Scatterplots based on the results from ten-fold cross validation for: (**a**) Predicted vs. observed precipitation monthly values; (**b**) Predicted vs. calculated *ED* monthly values. Black line symbolized the identity function *f(x)* = *x*.

3.3. Rainfall Erosivity and Its Uncertenty

The produced maps (Figures 5 and 6) illustrate the spatiotemporal distribution of *R* in Greece. The eastern, dryer part of the country has lower values than the wetter western part for the period during autumn and winter. During summer the convective activity over norther Greece produces higher *R* values than southern Greece, with the largest values occurring in the area of Thrace. The monthly temporal patterns illustrated in Figure 5, are compatible to the three areas that were identified in our previous study [30].

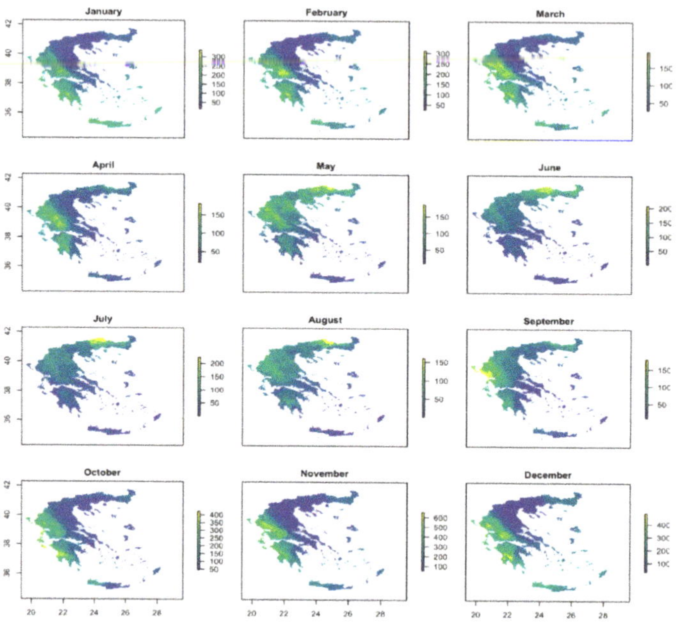

Figure 5. Monthly erosivity estimation. Units are in MJ·mm/ha/h/month.

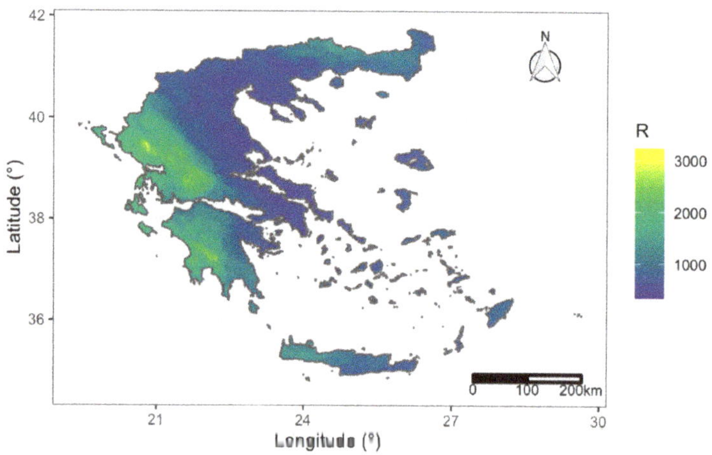

Figure 6. Annual rainfall erosivity. Values are in MJ·mm/ha/h/y. Available online in Supplementary Materials.

The lowest annual values of R were computed at the central mainland area of Greece (west and central Macedonia, Thessaly and Attica). The uncertainty map of annual *R* (Figure 7) follows the variation of the predicted annual erosivity and the scarcity of stations. The mean annual *R* (Table 7, Figure 6) has a value of 1039.0 MJ·mm/ha/h/y, with a range between 405.1 and 3160.2 MJ·mm/ha/h/y. The annual mean prediction error standard deviation is 116.9 MJ·mm/ha/h/y and its range is from 46.7 to 353.4 MJ·mm/ha/h/y (Table 7, Figure 7). The ratio of the mean annual error to the mean annual R is 11.25%, a value that was probably overestimated, given the cross-validation results of z-scores (Tables 5 and 6). These errors are not considered very high, bearing in mind the spatially sparse data and the data-sets limitations. Although these errors may affect the reliability of the assessment of the potential impacts of climate change, their effect is rather small when compared with the large uncertainties in GCMs projections in monthly and daily precipitation and rainfall intensity [84].

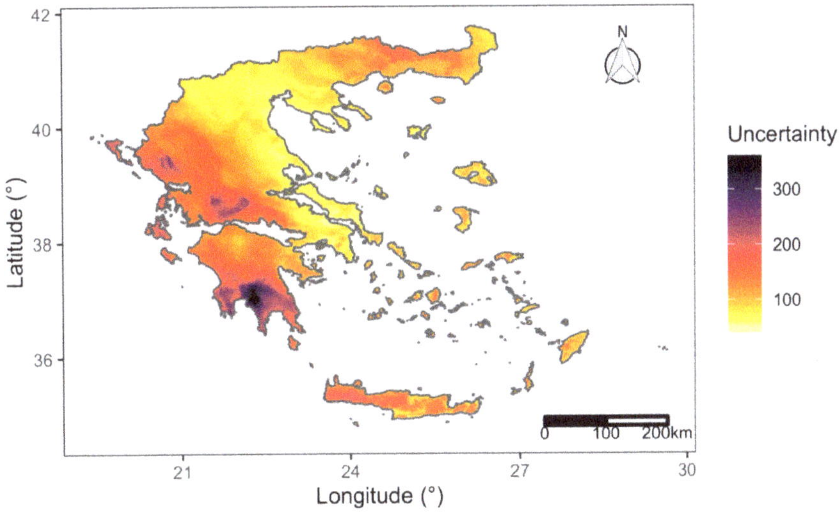

Figure 7. Annual rainfall erosivity prediction error standard deviation. Values are in MJ·mm/ha/h/y.

Table 7. Mean, minimum and maximum values of erosivity R and its uncertainty σ_R. Values are in MJ·mm/ha/h/month for monthly values and MJ·mm/ha/h/y for annual. Min is an abbreviation for minimum and max for maximum values.

	Jan	Feb	Mar	Apr	May	Jun	Jul	Aug	Sep	Oct	Nov	Dec	Annual
R_{min}	14.3	25.6	26.7	20.2	8.2	1.7	0.3	0.6	9.5	44.7	52.1	32.6	405.1
R	95.4	94.5	74.1	58.3	65.7	44.8	46.0	42.9	54.5	128.9	182.5	151.5	1039.0
R_{max}	369.6	314.3	234.7	193.5	187.2	207.4	222.6	161.5	188.8	441.7	739.2	505.7	3160.2
$\sigma_{R,min}$	10.7	9.0	8.2	3.9	4.0	2.2	4.5	1.2	4.7	11.2	18.7	10.8	46.7
σ_R	33.1	33.3	24.2	19.5	25.0	21.5	25.2	21.4	22.2	32.8	55.5	48.1	116.9
$\sigma_{R,max}$	97.6	129.3	74.8	61.2	62.0	69.8	100.9	83.4	50.2	95.6	220.5	192.9	353.4

The largest monthly mean R value are observed in November with 182.5 MJ·mm/ha/h/y and in December with 151.5 MJ·mm/ha/h/y and the lowest ones during summer months which have values about 44 MJ·mm/ha/h/y. The computed R values, due to the fact that R is linearly underestimated as the missing values ratio increases [30], were larger than the values reported by Panagos et al. [31] (i.e., the mean annual R was underestimated by 28.8%, the mean minimum by 381% and the maximum by 12%). On the contrary, the range of R values in this study is smaller (405.1–3160.2 MJ·mm/ha/h/y) than the respective values reported in [31] (was 84.2–2825 MJ·mm/ha/h/y).

In the above cited, older, study [31], the spatial distribution and annual values of R were affected by the used interpolated precipitation dataset. As a result, the highest values were calculated at the northwest corner of Greece. However, in this study, the maximum observed precipitation from HNMS's stations were recorded at the mountain range of Pindos, which is, also, the area with the maximum annual R values in our study, indicating that the two variables are consistent.

3.4. Projected Rainfall Erosivity Changes

In order to cope with the uncertainties in climatic models, five (5) GCMs-RCMs were evaluated in terms of $RMSE$, R^2 and CCC using the monthly precipitation maps that were created in this study. The EC-EARTH–KNMI-RACMO22E GCM-RCM (Table 8) gave the best results for each one of months and for all the error metrics. Given its performance, this GCM-RCM seemed to represent better monthly precipitation in Greece, for the historical time period and consequently used to estimate futures changes for the RCP forcing scenarios RCP4.5 and RCP8.5 and the years 2040, 2070 and 2100. However, the large errors presented in Table 8 indicate that statistical downscaling techniques have to be applied to the outputs of GCMs-RCMs for bias correction and reduction of errors.

Table 8. Mean monthly precipitation errors of the used GCMs-RCMs. *RMSE* units are in mm, R^2 and *CCC* are unitless.

	GCM	RCM	*RMSE*	R^2	*CCC*
1	EC-EARTH	DMI-HIRHAM5	64.4	−4.21	0.30
2	EC-EARTH	KNMI-RACMO22E	25.5	0.11	0.63
3	HadGEM2-ES	KNMI-RACMO22E	31.5	−0.37	0.57
4	MPI-ESM-LR	CSC-REMO2009	34.9	−0.47	0.50
5	NorESM1-M	DMI-HIRHAM5	61.9	−3.73	0.29

Using the RCP4.5 scenario, the ratio of the mean projected R to current values were −3.7%, −4.3% and −0.7% for 2040, 2070 and 2100, respectively. The same values for the RCP8.5, scenario were −9.0%, +0.01% and −14.6%. Annual R followed the annual ratio $\delta_f^{(s_i)}$ of projected to 30-years-mean historical precipitation, which fluctuated around zero for RCP4.5 (Figure 8a) and had a negative trend for the years 2040–2100 for RCP4.5.

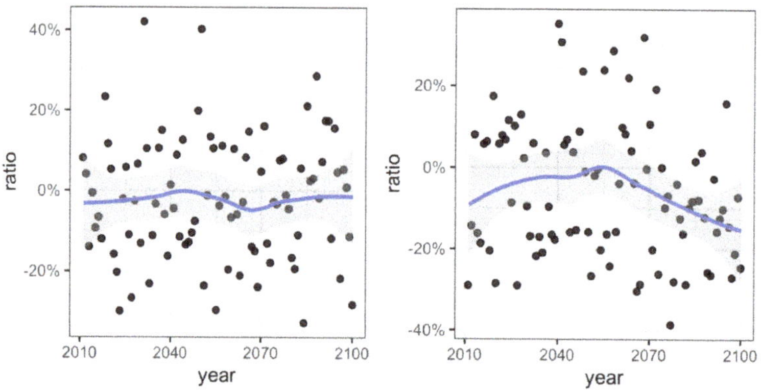

(a) (b)

Figure 8. Ratio δ_f^s of precipitation to the historical 30-years-mean annual on Greece for the scenarios: (a) Representative Concentration Pathway (RCP)45 and (b) RCP85. Values are unitless. With blue are marked the smooth lines and with grey bands the standard error variance produced by means of Local Polynomial Regression Fitting [81].

Despite the decrease of the projected R values, using RCP4.5, R, spatially, has a tendency to increase at the central parts of Macedonia and Thessaly that have the highest agricultural productivity in the country and the northern Thrace (Figure 9a). For RCP8.5 (Figure 9b) projected erosivity increases on 2070, having a hotspot at Thessaly and then decreases, following the precipitation trend (Figure 8b). In the previously reported study concerning Europe, that did not use information about rainfall intensities in the applied model, the projected change of R values for Greece on 2050 was +14.8% using the GCM HadGEM2 and RCP4.5 [10].

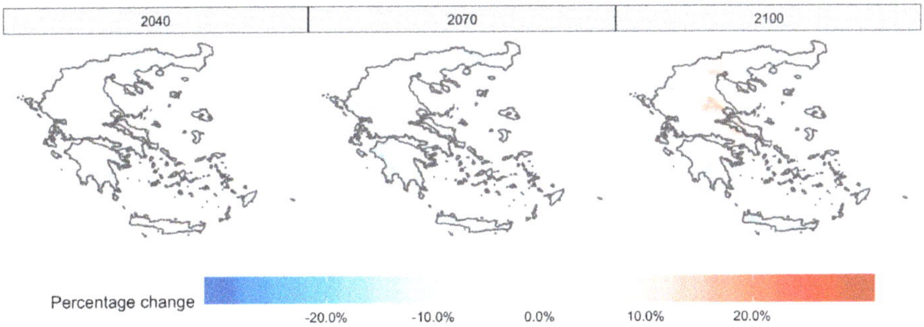

(a)

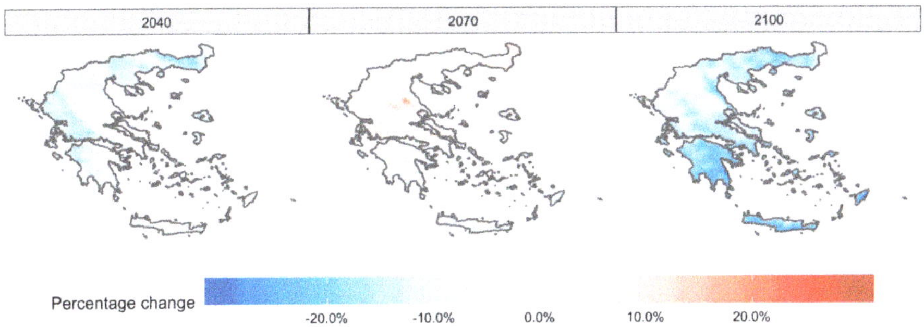

(b)

Figure 9. Projected percentage changes of rainfall erosivty, *R*, for the scenarios: (**a**) RCP4.5 and (**b**) RCP8.5 over the historical values.

The results of projected changes of *R* in both RCP scenarios are coherent to the precipitation trends that are reported by EURO-CORDEX, in which precipitation decreases by a larger magnitude in RCP8.5 than RCP4.5, from 1971–2000 to 2071–2100 [27]. This change in precipitation dominates in projected R's calculations and is depicted in Figure 9b for the year 2100.

4. Conclusions

The estimation of mean annual and monthly *R* values over Greece using precipitation records that suffered from a significant volume of missing values was the main result of this paper, utilizing as an intermediate step the creation of monthly precipitation and *ED* QRF models. The models of monthly precipitation had better performance than the ones of monthly *ED* in terms of prediction accuracy, mostly due to the scarcity of stations with calculated *ED* values. Validating the error of uncertainty reported from the models on cross validation locations, showed that the models, on average, overestimate the actual prediction error (i.e., the error of error predictions). More specifically, the findings of the present study can be summarized as follows:

1. The mean annual R in Greece is 1039 MJ·mm/ha/h/y, with a range between 405.1 and 3160.2 MJ·mm/ha/h/y, during the historical period 1971–2000. The highest values are calculated at the mountain range of Pindos and the lowest at central Greece's mainland.

2. The calculated monthly mean *R* values follow the already documented spatiotemporal characteristics of precipitation depth and intensity over the country.
3. The climatic model EC-EARTH-KNMI-RACMO22E from the Royal Netherlands Meteorological Institute better reproduces the monthly precipitation for the historical period 1971–2000 in Greece than the other four GCMs-RCMs used and tested in this study.
4. The projected mean annual erosivity, *R*, as an average over Greece, follows, in general, the projected changes of precipitation from the selected GCM-RCM model but not in a spatially homogenous way.

The results about future values of *R* inherit a set of uncertainties that have to do with the limitations of climatic models in general and the assumptions about future temporal stationarity of *ED* that had been made in our study, based on the observed stationarity of *ED* for the historical time. This holds true, as well as for the calculated current *ED* values due to missing values from the utilized timeseries. Future research will eventually provide more robust climatic models, as computational power increases and research continues, and hopefully we will also have high quality, high density, observed precipitation data for longer durations and more stations to estimate more accurately current and projected rainfall erosivity in the country.

Supplementary Materials: The annual *R* values raster is available online at: https://doi.org/10.5281/zenodo.3692645.

Author Contributions: K.V. designed the study, developed the coding and performed the analysis, E.S. and A.L. organized and wrote the manuscript. All authors have read and agreed to the published version of the manuscript.

Funding: This research received no external funding

Acknowledgments: The data importing, analysis and presentation were done using the open source *R* language for statistical computing and graphics [82] using the packages: hydroscoper [52], hyetor [85], ranger [67], tuneRanger [83], raster [86] and ggplot2 [87].

Conflicts of Interest: The authors declare no conflict of interest.

References

1. Nearing, M.A.; Yin, S.; Borrelli, P.; Polyakov, V.O. Rainfall erosivity: An historical review. *Catena* **2017**, *157*, 357–362. [CrossRef]
2. Geeson, N.A.; Brandt, C.J.; Thornes, J.B. *Mediterranean Desertification: A Mosaic of Processes and Responses*; John Wiley & Sons: West Sussex, UK, 2003.
3. Pruski, F.F.; Nearing, M.A. Runoff and soil-Loss responses to changes in precipitation: A computer simulation study. *J. Soil Water Conserv.* **2002**, *57*, 7–16.
4. Nearing, M.A.; Pruski, F.F.; O'neal, M.R. Expected climate change impacts on soil erosion rates: A review. *J. Soil Water Conserv.* **2004**, *59*, 43–50.
5. Kosmas, C.; Tsara, M.; Moustakas, N.; Karavitis, C. Identification of indicators for desertification. *Ann. Arid Zone* **2003**, *42*, 393–416.
6. Salvati, L.; Carlucci, M. The impact of mediterranean land degradation on agricultural income: A short-Term scenario. *Land Use Policy* **2013**, *32*, 302–308. [CrossRef]
7. Webb, N.P.; Marshall, N.A.; Stringer, L.C.; Reed, M.S.; Chappell, A.; Herrick, J.E. Land degradation and climate change: Building climate resilience in agriculture. *Front. Ecol. Environ.* **2017**, *15*, 450–459. [CrossRef]
8. Griggs, D.; Stafford-Smith, M.; Gaffney, O.; Rockström, J.; Öhman, M.C.; Shyamsundar, P.; Steffen, W.; Glaser, G.; Kanie, N.; Noble, I. Sustainable development goals for people and planet. *Nature* **2013**, *495*, 305–307. [CrossRef]
9. Keesstra, S.D.; Bouma, J.; Wallinga, J.; Tittonell, P.; Smith, P.; Cerdà, A.; Montanarella, L.; Quinton, J.N.; Pachepsky, Y.; van der Putten, W.H.; et al. The significance of soils and soil science towards realization of the United Nations Sustainable Development Goals. *Soil* **2016**, *2*, 111–128. [CrossRef]
10. Panagos, P.; Ballabio, C.; Meusburger, K.; Spinoni, J.; Alewell, C.; Borrelli, P. Towards estimates of future rainfall erosivity in Europe based on REDES and WorldClim datasets. *J. Hydrol.* **2017**, *548*, 251–262. [CrossRef]

11. Wischmeier, W.H.; Smith, D.D. *Predicting Rainfall Erosion Losses-A Guide to Conservation Planning*; USDA, Agriculture Handbook No. 537; Government Printing Office: Washington DC, USA, 1978.
12. Kinnell, P.I.A. Event soil loss, runoff and the Universal Soil Loss Equation family of models: A review. *J. Hydrol.* **2010**, *385*, 384–397. [CrossRef]
13. Renard, K.G.; Freimund, J.R. Using monthly precipitation data to estimate the R-Factor in the revised USLE. *J. Hydrol.* **1994**, *157*, 287–306. [CrossRef]
14. Vantas, K.; Sidiropoulos, E.; Evangelides, C. Rainfall erosivity and its estimation: Conventional and machine learning methods. In *Soil Erosion-Rainfall Erosivity and Risk Assessment*; IntechOpen: London, UK, 2019.
15. USDA-ARS. *Science Documentation: Revised Universal Soil Loss Equation*; Version 2 (RUSLE 2); USDA-Agricultural Research Service: Washington, DC, USA, 2013.
16. Feidas, H.; Noulopoulou, C.; Makrogiannis, T.; Bora-Senta, E. Trend analysis of precipitation time series in Greece and their relationship with circulation using surface and satellite data: 1955–2001. *Theor. Appl. Climatol.* **2007**, *87*, 155–177. [CrossRef]
17. Feidas, H.; Karagiannidis, A.; Keppas, S.; Vaitis, M.; Kontos, T.; Zanis, P.; Melas, D.; Anadranistakis, E. Modeling and mapping temperature and precipitation climate data in Greece using topographical and geographical parameters. *Theor. Appl. Climatol.* **2014**, *118*, 133–146. [CrossRef]
18. Bartzokas, A.; Lolis, C.J.; Metaxas, D.A. A study on the intra-Annual variation and the spatial distribution of precipitation amount and duration over Greece on a 10 day basis. *Int. J. Climatol. J. R. Meteorol. Soc.* **2003**, *23*, 207–222. [CrossRef]
19. Xoplaki, E.; Luterbacher, J.; Burkard, R.; Patrikas, I.; Maheras, P. Connection between the large-Scale 500 hPa geopotential height fields and precipitation over Greece during wintertime. *Clim. Res.* **2000**, *14*, 129–146. [CrossRef]
20. Tolika, K.; Maheras, P. Spatial and temporal characteristics of wet spells in Greece. *Theor. Appl. Climatol.* **2005**, *81*, 71–85. [CrossRef]
21. Maheras, P.; Anagnostopoulou, C. Circulation Types and Their Influence on the Interannual Variability and Precipitation Changes in Greece. In *Mediterranean Climate*; Bolle, H.-J., Ed.; Springer: Berlin/Heidelberg, Germany, 2003; pp. 215–239. ISBN 978-3-642-62862-7.
22. Maheras, P.; Tolika, K.; Anagnostopoulou, C.; Vafiadis, M.; Patrikas, I.; Flocas, H. On the relationships between circulation types and changes in rainfall variability in Greece. *Int. J. Climatol. J. R. Meteorol. Soc.* **2004**, *24*, 1695–1712. [CrossRef]
23. Gofa, F.; Mamara, A.; Anadranistakis, M.; Flocas, H. Developing Gridded Climate Data Sets of Precipitation for Greece Based on Homogenized Time Series. *Climate* **2019**, *7*, 68. [CrossRef]
24. McGuffie, K.; Henderson-Sellers, A. *The Climate Modelling Primer*; John Wiley & Sons: West Sussex, UK, 2014.
25. Meinshausen, M.; Smith, S.J.; Calvin, K.; Daniel, J.S.; Kainuma, M.L.T.; Lamarque, J.-F.; Matsumoto, K.; Montzka, S.A.; Raper, S.C.B.; Riahi, K.; et al. The RCP greenhouse gas concentrations and their extensions from 1765 to 2300. *Clim. Chang.* **2011**, *109*, 213–241. [CrossRef]
26. Giorgi, F. Thirty Years of Regional Climate Modeling: Where Are We and Where Are We Going next? *J. Geophys. Res. Atmos.* **2019**, *124*, 5696–5723. [CrossRef]
27. Jacob, D.; Petersen, J.; Eggert, B.; Alias, A.; Christensen, O.B.; Bouwer, L.M.; Braun, A.; Colette, A.; Déqué, M.; Georgievski, G. EURO-CORDEX: New high-Resolution climate change projections for European impact research. *Reg. Environ. Chang.* **2014**, *14*, 563–578. [CrossRef]
28. European Environment Agency. *Climate Change. Impacts and Vulnerability in Europe 2016: An Indicator-Based Report*; European Environment Agency: Luxembourg, 2017; ISBN 978-92-9213-835-6.
29. Vantas, K.; Sidiropoulos, E.; Loukas, A. Temporal and Elevation Trend Detection of Rainfall Erosivity Density in Greece. *Proceedings* **2018**, *7*, 10. [CrossRef]
30. Vantas, K.; Sidiropoulos, E.; Loukas, A. Robustness Spatiotemporal Clustering and Trend Detection of Rainfall Erosivity Density in Greece. *Water* **2019**, *11*, 1050. [CrossRef]
31. Panagos, P.; Ballabio, C.; Borrelli, P.; Meusburger, K. Spatio-Temporal analysis of rainfall erosivity and erosivity density in Greece. *Catena* **2016**, *137*, 161–172. [CrossRef]
32. Diodato, N.; Bellocchi, G. MedREM, a rainfall erosivity model for the Mediterranean region. *J. Hydrol.* **2010**, *387*, 119–127. [CrossRef]
33. Hijmans, R.J.; Cameron, S.E.; Parra, J.L.; Jones, P.G.; Jarvis, A. Very high resolution interpolated climate surfaces for global land areas. *Int. J. Climatol.* **2005**, *25*, 1965–1978. [CrossRef]

34. Fiener, P.; Neuhaus, P.; Botschek, J. Long-Term trends in rainfall erosivity–Analysis of high resolution precipitation time series (1937–2007) from Western Germany. *Agric. For. Meteorol.* **2013**, *171*, 115–123. [CrossRef]
35. Verstraeten, G.; Poesen, J.; Demarée, G.; Salles, C. Long-Term (105 years) variability in rain erosivity as derived from 10-min rainfall depth data for Ukkel (Brussels, Belgium): Implications for assessing soil erosion rates. *J. Geophys. Res.* **2006**, *111*. [CrossRef]
36. Hanel, M.; Pavlásková, A.; Kyselý, J. Trends in characteristics of sub-daily heavy precipitation and rainfall erosivity in the Czech Republic. *Int. J. Climatol.* **2016**, *36*, 1833–1845. [CrossRef]
37. Amanambu, A.C.; Li, L.; Egbinola, C.N.; Obarein, O.A.; Mupenzi, C.; Chen, D. Spatio-Temporal variation in rainfall-Runoff erosivity due to climate change in the Lower Niger Basin, West Africa. *Catena* **2019**, *172*, 324–334. [CrossRef]
38. Duulatov, E.; Chen, X.; Amanambu, A.; Ochege, U.F.; Orozbaev, R.; Issanova, G.; Omurakunova, G. Projected Rainfall Erosivity Over Central Asia Based on CMIP5 Climate Models. *Water* **2019**, *11*, 897. [CrossRef]
39. Almagro, A.; Oliveira, P.T.S.; Nearing, M.A.; Hagemann, S. Projected climate change impacts in rainfall erosivity over Brazil. *Sci. Rep.* **2017**, *7*, 8130. [CrossRef] [PubMed]
40. Plangoen, P.; Udmale, P. Impacts of Climate Change on Rainfall Erosivity in the Huai Luang Watershed, Thailand. *Atmosphere* **2017**, *8*, 143. [CrossRef]
41. Zhu, Q.; Yang, X.; Ji, F.; Liu, D.L.; Yu, Q. Extreme rainfall, rainfall erosivity, and hillslope erosion in Australian Alpine region and their future changes. *Int. J. Climatol.* **2019**. [CrossRef]
42. Gericke, A.; Kiesel, J.; Deumlich, D.; Venohr, M. Recent and Future Changes in Rainfall Erosivity and Implications for the Soil Erosion Risk in Brandenburg, NE Germany. *Water* **2019**, *11*, 904. [CrossRef]
43. Zhang, Y.-G.; Nearing, M.A.; Zhang, X.-C.; Xie, Y.; Wei, H. Projected rainfall erosivity changes under climate change from multimodel and multiscenario projections in Northeast China. *J. Hydrol.* **2010**, *384*, 97–106. [CrossRef]
44. Breiman, L. Random forests. *Mach. Learn.* **2001**, *45*, 5–32. [CrossRef]
45. Meinshausen, N. Quantile Regression Forests. *J. Mach. Learn. Res.* **2006**, *7*, 983–999.
46. Nussbaum, M.; Spiess, K.; Baltensweiler, A.; Grob, U.; Keller, A.; Greiner, L.; Schaepman, M.E.; Papritz, A.J. Evaluation of digital soil mapping approaches with large sets of environmental covariates. *Soil* **2018**, *4*, 1–22. [CrossRef]
47. da Silva Júnior, J.C.; Medeiros, V.; Garrozi, C.; Montenegro, A.; Gonçalves, G.E. Random forest techniques for spatial interpolation of evapotranspiration data from Brazilian's Northeast. *Comput. Electron. Agric.* **2019**, *166*, 105017. [CrossRef]
48. Li, J.; Heap, A.D.; Potter, A.; Daniell, J.J. Application of machine learning methods to spatial interpolation of environmental variables. *Environ. Model. Softw.* **2011**, *26*, 1647–1659. [CrossRef]
49. Hengl, T.; Walsh, M.G.; Sanderman, J.; Wheeler, I.; Harrison, S.P.; Prentice, I.C. Global mapping of potential natural vegetation: An assessment of machine learning algorithms for estimating land potential. *PeerJ* **2018**, *6*, e5457. [CrossRef] [PubMed]
50. Fouedjio, F.; Klump, J. Exploring prediction uncertainty of spatial data in geostatistical and machine learning approaches. *Environ. Earth Sci.* **2019**, *78*, 38. [CrossRef]
51. Hengl, T.; Nussbaum, M.; Wright, M.N.; Heuvelink, G.B.M.; Gräler, B. Random forest as a generic framework for predictive modeling of spatial and spatio-Temporal variables. *PeerJ* **2018**, *6*, e5518. [CrossRef]
52. Vantas, K. hydroscoper: R interface to the Greek National Data Bank for Hydrological and Meteorological Information. *J. Open Source Softw.* **2018**, *3*, 625. [CrossRef]
53. Hellenic National Meteorological Service. Available online: http://emy.gr/emy/el/ (accessed on 14 January 2020).
54. Copernicus Climate Change Service. Available online: https://cds.climate.copernicus.eu/cdsapp#!/home (accessed on 14 January 2020).
55. Schulzweida, U. CDO User Guide. Available online: http://doi.org/10.5281/zenodo.3539275 (accessed on 14 January 2020).
56. Farr, T.G.; Rosen, P.A.; Caro, E.; Crippen, R.; Duren, R.; Hensley, S.; Kobrick, M.; Paller, M.; Rodriguez, E.; Roth, L.; et al. The Shuttle Radar Topography Mission. *Rev. Geophys.* **2007**, *45*. [CrossRef]
57. McGregor, K.C.; Bingner, R.L.; Bowie, A.J.; Foster, G.R. Erosivity index values for northern Mississippi. *Trans. ASAE* **1995**, *38*, 1039–1047. [CrossRef]

58. Mineo, C.; Ridolfi, E.; Moccia, B.; Russo, F.; Napolitano, F. Assessment of Rainfall Kinetic-Energy–Intensity Relationships. *Water* **2019**, *11*, 1994. [CrossRef]
59. Panagos, P.; Borrelli, P.; Spinoni, J.; Ballabio, C.; Meusburger, K.; Beguería, S.; Klik, A.; Michaelides, S.; Petan, S.; Hrabalíková, M.; et al. Monthly Rainfall Erosivity: Conversion Factors for Different Time Resolutions and Regional Assessments. *Water* **2016**, *8*, 119. [CrossRef]
60. Yin, S.; Xie, Y.; Nearing, M.A.; Wang, C. Estimation of rainfall erosivity using 5-to 60-Min fixed-interval rainfall data from China. *Catena* **2007**, *70*, 306–312. [CrossRef]
61. Biau, G.; Scornet, E. A random forest guided tour. *TEST* **2016**, *25*, 197–227.
62. Roy, M.-H.; Larocque, D. Robustness of random forests for regression. *J. Nonparametric Stat.* **2012**, *24*, 993–1006. [CrossRef]
63. Kuhn, M.; Johnson, K. *Applied Predictive Modeling*; Springer: New York, NY, USA, 2013; ISBN 978-1-4614-6848-6.
64. Genuer, R.; Poggi, J.-M.; Tuleau-Malot, C. Variable selection using random forests. *Pattern Recognit. Lett.* **2010**, *31*, 2225–2236. [CrossRef]
65. Díaz-Uriarte, R.; De Andres, S.A. Gene selection and classification of microarray data using random forest. *BMC Bioinform.* **2006**, *7*, 3. [CrossRef] [PubMed]
66. Liaw, A.; Wiener, M. Classification and regression by randomForest. *R News* **2002**, *2*, 18–22.
67. Wright, M.N.; Ziegler, A. A fast implementation of random forests for high dimensional data in C++ and R. *J. Stat. Softw.* **2017**, *77*, 1–17. [CrossRef]
68. Breiman, L. *Classification and Regression Trees*; CRC Press: New York, NY, USA, 2017.
69. Efron, B. Bootstrap methods: Another look at the jackknife. In *Breakthroughs in Statistics*; Springer: Berlin/Heidelberg, Germany, 1992; pp. 569–593.
70. Hastie, T.; Tibshirani, R.; Friedman, J.H. *The Elements of Statistical Learning: Data Mining, Inference, and Prediction*, 2nd ed.; Springer Series in Statistics; Springer: New York, NY, USA, 2009; ISBN 978-0-387-84857-0.
71. James, G.; Witten, D.; Hastie, T.; Tibshirani, R. *An Introduction to Statistical Learning*; Springer: Berlin/Heidelberg, Germany, 2013; Volume 112.
72. Stekhoven, D.J.; Buhlmann, P. MissForest–Non-Parametric missing value imputation for mixed-Type data. *Bioinformatics* **2012**, *28*, 112–118. [CrossRef]
73. Kvålseth, T.O. Cautionary note about r2. *Am. Stat.* **1985**, *39*, 279–285.
74. Lin, L.I.-K. A concordance correlation coefficient to evaluate reproducibility. *Biometrics* **1989**, 255–268. [CrossRef]
75. Liao, J.J.; Lewis, J.W. A note on concordance correlation coefficient. *PDA J. Pharm. Sci. Technol.* **2000**, *54*, 23–26.
76. Bivand, R.; Pebesma, E.J.; Gómez-Rubio, V. *Applied Spatial Data Analysis with R*, 2nd ed.; Use R! Springer: New York, NY, USA, 2013; ISBN 978-1-4614-7617-7.
77. Barford, N.C. *Experimental Measurements: Precision, Error and Truth*; Wiley: New York, NY, USA, 1985.
78. Wischmeier, W.H.; Smith, D.D. Rainfall energy and its relationship to soil loss. *Trans. Am. Geophys. Union* **1958**, *39*, 285. [CrossRef]
79. Foster, G.R.; Lombardi, F.; Moldenhauer, W.C. Evaluation of Rainfall-Runoff Erosivity Factors for Individual Storms. *Trans. ASAE* **1982**, *25*, 0124–0129. [CrossRef]
80. Hatzianastassiou, N.; Katsoulis, B.; Pnevmatikos, J.; Antakis, V. Spatial and temporal variation of precipitation in Greece and surrounding regions based on global precipitation climatology project data. *J. Clim.* **2008**, *21*, 1349–1370. [CrossRef]
81. Shyu, W.M.; Grosse, E.; Cleveland, W.S. Local regression models. In *Statistical Models in S*; Chapman and Hall/CRC: Boca Raton, FL, USA, 2017; pp. 309–376. ISBN 978-0-412-83040-2.
82. R Core Team. *R. A Language and Environment for Statistical Computing*; R Core Team: Vienna, Austria, 2019.
83. Probst, P.; Wright, M.; Boulesteix, A. L. Hyperparameters and tuning strategies for random forest. *Wiley Interdiscip. Rev. Data Min. Knowl. Discov.* **2018**. [CrossRef]
84. Sunyer, M.A.; Hundecha, Y.; Lawrence, D.; Madsen, H.; Willems, P.; Martinkova, M.; Vormoor, K.; Bürger, G.; Hanel, M.; Kriaučiūnienė, J.; et al. Inter-Comparison of statistical downscaling methods for projection of extreme precipitation in Europe. *Hydrol. Earth Syst. Sci.* **2015**, *19*, 1827–1847. [CrossRef]
85. Vantas, K. hyetor: R Package to Analyze Fixed Interval Precipitation Time Series. Available online: https://github.com/kvantas/hyetor (accessed on 14 January 2020).

86. Hijmans, R.J. Raster: Geographic Data Analysis and Modelling. Available online: https://github.com/rspatial/raster (accessed on 14 January 2020).
87. Wickham, H. *ggplot2: Elegant Graphics for Data Analysis*; Springer: New York, NY, USA, 2016.

water

MDPI

Article

Evaluation of Hydromulches as an Erosion Control Measure Using Laboratory-Scale Experiments

Matthew D. Ricks [1,2,*], Wesley T. Wilson [3], Wesley C. Zech [4], Xing Fang [1] and Wesley N. Donald [1]

1 Department of Civil Engineering, Auburn University, Auburn, AL 36849-5337, USA; xing.fang@auburn.edu (X.F.); donalwn@auburn.edu (W.N.D.)
2 Liberty Utilities, 2300 Victory Drive, Columbus, GA 31902-3455 USA
3 HIIT, 216 Seven Farms Drive, Charleston, SC 29492-7971, USA; wwilson@hitt-gc.com
4 Department of Civil, Construction, and Environmental Engineering, The University of Alabama at Birmingham, Birmingham, AL 35294-4440, USA; zechwes@uab.edu
* Correspondence: mzr0062@auburn.edu; Tel.: +1-706-566-9798

Received: 28 December 2019; Accepted: 8 February 2020; Published: 13 February 2020

check for updates

Abstract: Discharge of sediment-laden stormwater from active construction sites, such as highway construction projects, continues to be a growing concern in the construction industry. Therefore, there has been an increased interest in research efforts to test many different erosion and sediment control practices. The purpose of this research effort was to test the laboratory-scale performance of four hydromulches and two methods of mulching (crimped and tackified), normalized to a bare soil control condition using 0.6 m (2 ft) wide by 1.2 m (4 ft) long test plots. The treatments consisted of a (1) bare soil control, (2) conventional straw, crimped, (3) conventional straw, tackified, (4) wood fiber hydromulch, (5) straw and cotton hydromulch, (6) cotton fiber reinforced matrix hydromulch, and (7) bonded wheat fiber matrix hydromulch. Each treatment was subject to simulated rainfall, divided into four 15 min rainfall events with 15 min breaks in between, producing a total cumulative rainfall of 11.2 cm (4.4 in.). To determine the overall performance of each treatment, turbidity and soil loss measurements were continuously collected from plot runoff. The products tested provided a reduction in turbidity of 80%, 98%, 85%, 92%, 95%, and 99%; and a soil loss reduction of 96%, 98%, 94%, 97%, 99%, and 100%, respectively. Overall, the results showed that the four tested hydromulch practices and conventional straw applications were successful in controlling and reducing erosion under laboratory-scale simulated rainfall conditions.

Keywords: erosion control; laboratory-scale testing; simulated rainfall; runoff

1. Introduction

The discharge of sediment-laden stormwater from active construction sites, such as highway construction projects, is a growing concern in the construction industry [1]. The United States Environmental Protection Agency (USEPA) labels such discharge as nonpoint source (NPS) pollution, which is defined as land runoff, precipitation, atmospheric deposition, seepage, or hydrologic modification that does not meet the legal definition of 'point source' in Section 502 (14) of the Clean Water Act [2].

Soil erosion is considered the largest contributor to NPS pollution in the U.S. [3]. Construction sites are known to be a significant contributor to soil erosion by exhibiting soil loss rates that are 20 times greater from construction sites than agricultural lands, and 1000 to 2000 times greater than forest lands [4,5]. Studies have shown that erosion rates on cut slopes of roadways has varied from 5.93 mm/ha (0.09 in./ac. or in./ac.) up to 70 mm/ha (1.12 in./ac.) [6]. When soil is eroded from construction sites, other harmful particulates such as fertilizers, pesticides, metals, and fuels attach to the soil and are transported into municipal separate storm sewer systems (MS4s) [7,8]. Polluted MS4s transport construction site runoff directly to surface waters, ultimately causing sedimentation. In the U.S. alone, "sedimentation impairs 84,503 river and stream miles (12% of the assessed river and stream miles and 31% of the impaired river and stream miles)" [9]. Sedimentation of surface water can lead to deterioration of aquatic habitats, rapid loss of storage capacity of reservoirs, eroded streambanks, and increased turbidity of the waters thereby reducing photosynthesis, and clogging fish gills [10]. An annual estimate of $17 billion is spent in the U.S. alone in an effort to control onsite sedimentation, bringing the national total to nearly $60 billion in erosion and sediment control activities [11]. Thus, the combination of environmental and economic downfalls related to erosion and sedimentation in the construction industry has developed a need for scientific research to be performed to understand the overall performance of erosion and sediment control (ESC) practices used at the federal, state, and local levels.

Within the construction industry, there are numerous types of erosion controls. The focus of this research effort is to test the performance of the following surface cover treatments: (1) conventional straw, crimped, (2) conventional straw, tackified, (3) wood fiber hydromulch (HM) (Excel® Fibermulch II), (4) straw and cotton hydromulch (Geoskin®), (5) cotton fiber reinforced matrix hydromulch (FRM) (HydraCX2®), and (6) bonded wheat fiber matrix hydromulch (FM) (Hydrostraw® BFM).

1.1. Mulching as an Erosion Control

Mulching is defined as an erosion control practice that uses materials such as shredded paper, grass, hay, wood chips, wood fibers, straw, or gravel to stabilize exposed or recently planted soil surfaces [12,13]. Surface mulch has been found to be one of the most effective, practical means of controlling runoff and erosion on disturbed land prior to vegetation establishment; however it is most effective when used in conjunction with vegetation [12,14,15]. Researchers [16–19] have reported that mulches used to control erosion have a two-fold advantage: (1) reduce soil loss and (2) protect grass seeds and soil amendments from being washed away. Additionally, mulches are capable of reducing solar radiation, suppressing fluctuations of soil temperature, reducing water loss through evaporation, increases interception storage capacity, dissipating the kinetic energy from the raindrops impact, and helping to prevent soil crust formation [17,18,20–23]. Research has also shown that mulching can reduce sediment yields by over 80% when applied at a rate of 2000 kg/ha (1784 lb./ac.) [23,24].

The purpose of testing conventional straw was to have a traditional, low-cost, widely used erosion control practice to compare to the performance of hydromulch products. Straw is one of the most widely used ground covers used to reduce erosion on construction sites [25], and has been reported to reduce erosion rates by more than 90% if applied at sufficient rates [22,26–28]. Turgeon [21] states that straw is also capable of encouraging grass establishment by reducing runoff, increasing infiltration, and improving soil conditions.

Straw crimpers are typically used to crimp or punch straw into the soil when the soil is not too sandy [29]. If crimpers are not available or necessary, liquid mulch binders are used to 'tack' mulch by spraying the tack on top of the straw [15].

There are advantages and disadvantages to using straw mulch for erosion control. The advantages are that it is inexpensive, quick, and easy to apply using a straw-blower, capable of achieving efficient grass growth, and water is not needed for application. Straw mulch has also been found to perform as well as or better than hydromulch products when applied at sufficient rates [30]. Other studies have shown straw mulch to not only reduce soil erosion in the short term, but also by aiding in vegetation establishment through the long-term reduction of soil erosion [31]. Conversely, disadvantages of conventional straw include that it does not prevent soil loss as well as more expensive erosion products (e.g., erosion control blankets, compost, etc.), is susceptible to wind if not properly anchored, may introduce weed seeds, and fines from straw blowers can drift long distances [29].

1.2. Hydraulically Applied Mulch (As Known as Hydromulch)

Hydraulically applied mulches, referred to herein as 'hydromulches', have shown continuous evolution and improvement over the past 50 years. Advancements in technology have resulted in the production of equipment and materials that offer enhanced performance and greater productivity over many traditional methods of erosion control. Hydromulch has been shown to meet the required planting depth for small seeded species [32]. In other studies, hydromulch has been shown to reduce the sediment yield by about 75% when compared to bare plots [33]. There is a knowledge gap between the cost-effectiveness and performance benefits of new products [18,34–36] such as hydromulches, largely due to newly evolving technologies as well as a lack of research involving hydromulch products.

The introduction of water, refined fiber matrices, tackifiers, super-absorbents, flocculating agents, man-made fibers, plant biostimulants, and other performance enhancing additives to hydromulching practices on slopes has forced federal, state, and local governments to develop hydromulch guidelines. ASTM International (ASTM) has proposed new standards for testing hydraulically applied erosion control products (HECPs). Additionally, the Erosion Control Technology Council (ECTC) has divided HECPs into five distinct categories, relevant to their corresponding functional longevity, erosion control effectiveness, and vegetative establishment [29,37]. Specific to this study, the addition of a tackifier to a hydromulch has been shown to increase the effectiveness of the hydromulch as a soil cover due to the tackifier bonding with the soil particles and creating a more hydrophobic environment [38]. Prats et al. [23] determined that the initial reduction in soil erosion on a plot treated with hydromulch was attributed to the initial protective cover provided by the mulch to minimize splash erosion.

McLaughlin and Brown [27] conducted large- and laboratory-scale tests on four ground cover practices: straw mulch, straw erosion control blanket, wood fiber, and a mechanically bonded fiber matrix (MBFM) hydromulch. In their study, it was reported that the ground covers reduced runoff turbidity by a factor of four or greater when compared to bare soil. More specifically, on the controlled, laboratory-scale tests, the MBFM reduced average turbidity by approximately 85% and sediment loss by about 86% in comparison to a bare soil control.

Holt et al. [39] performed laboratory-scale tests on six hydromulch treatments using 0.6 m (2 ft) wide by 3.05 m (10 ft) long by 7.62 cm (3 in.) deep trays at a 15.7% slope. The following six hydromulches were applied by hand at 1120 kg/ha (1000 lb./ac.) and 2240 kg/ha (2000 lb./ac.): wood hydromulch, paper hydromulch, cottonseed hulls hydromulch, cotton byproduct (COBY) hydromulch produced from stripper waste (COBY Red), COBY produced from picker waste (COBY Yellow), and COBY produced from ground stripper waste (COBY Green). COBY is a term used in Holt's report to represent a patented cotton by product of cottonseed hulls [40]. The respective soil treatments with an application rate of 1120 kg/ha (1000 lb./ac.) achieved soil loss reductions of 35%, 58%, 84%, 90%, 80%, and 80% for wood, paper, cotton-seed hulls, COBY red, COBY yellow, and COBY green. When the application rate was increased to 2240 kg/ha (2000 lb./ac.), the respective soil treatments achieved soil loss reductions of 19%, 32%, 79%, 88%, 88%, and 68% for wood, paper, cotton-seed hulls, COBY red, COBY yellow, and COBY green.

In 2002, Landloch [41] studied the performance of four hydromulch treatments using 15 plots that were 5 m long by 1.5 m wide (16.4 ft long by 4.9 ft wide) at a 25% slope. The four hydromulches tested were paper hydromulch, flax hydromulch, flax plus paper hydromulch, and sugar cane hydromulch, applied at a rate of 1000 (893 lb./ac.), 2500 (2232 lb./ac.), 3250 (2900 lb./ac.), and 5000 kg/ha (4464 lb./ac.), respectively. The respective treatments achieved soil loss reductions of 80%, 85%, 96%, and 96% for paper, flax, flax plus paper, and sugar cane.

Benik et al. [42] developed a study comparing the effectiveness of five treatments, including Soil Guard® which is a bonded fiber matrix (BFM). In their experiments, the BFM was applied at a minimum rate of 3360 kg/ha (3000 lb./ac.). The BFM reduced average sediment yield by approximately 94%.

Buxton and Caruccio [43] evaluated 19 soil stabilizing and erosion control treatments, four of them were hydromulches without tackifiers. The plot sizes used were approximately 1.5 m (5 ft) wide by 3 m (10 ft) long at a 12% to 15% slope. The four hydromulches tested were Conwed wood fiber mulch,

Superior wood fiber mulch, Silva wood fiber mulch, and Pulch; each hydromulch was applied at a rate of 1344 kg/ha (1200 lb./ac.). In the study of Buxton and Caruccio [43], effectiveness of the hydromulches were measured using a vegetative maintenance (VM) and erosion control value, which in 1979 was a new parameter in the Universal Soil Loss Equation (USLE), and represented total loss ratio expressed as a decimal. These values ranged from 0.0 to 1.0, where a value of 1.0 means the erosion control practice had no effect in reducing erosion. The VM values for Buxton and Cauccio's [43] report were translated below in Table 1 to measure erosion control performance in soil loss reduction percentage.

Table 1. Summary of reviewed hydromulch practices.

Study	Type of Hydromulch	Test Scale	Slope	Application Rate (kg/ha)	Soil Loss Reduction (%)
McLaughlin and Brown [27]	MBFM	Large and laboratory	10% and 20%	3360	86
Holt et al. [39]	Wood	Laboratory	15.7%	1120	35
	Paper				58
	Cotton-seed hulls				84
	COBY red				90
	COBY yellow				80
	COBY green				80
	Wood			2240	19
	Paper				32
	Cotton-seed hulls				79
	COBY red				88
	COBY yellow				88
	COBY green				68
Benik et al. [42]	BFM	Large	35%	3360	94
Landloch [41]	Paper	Large	25%	1000	80
	Flax			2500	85
	Flax plus paper			3250	96
	Sugar Cane			5000	96
Buxton and Caruccio [43]	Conwed *	Large	12% to 15%	1344	77
	Superior *				73
	Silva *				35
	Pulch *				72
Babcock and McLaughlin [25]	Wood	Laboratory	33%	1970	19
				2940	8
Robichaud et al. [44]	Wood	Large	Various	1100	65
				600	19

* All are wood-fiber hydromulches.

Babcock and McLaughlin [25] evaluated straw mulch, with and without polyacrylamide (PAM), and a wood fiber hydromulch, with and without PAM, on the effectiveness of reducing erosion and improving the water quality of the runoff. The plot sizes used were 1 m by 2 m (3.3 ft by 6.6 ft) on a ~33% slope. The plots were subjected to a total rainfall of 3.05 cm (1.2 in.) at an intensity of 3.7 cm/h (1.5 in./h). The mulch was applied at a rate of 2240 kg/ha (1998 lb./ac.), while the hydromulch was applied at two separate application rates: 1970 kg/ha (1758 lb./ac.) and 2940 kg/ha (2623 lb./ac.). This study found that hydromulch applied at a rate of 2940 kg/ha (2623 lb./ac.) provided a soil loss reduction of 8% and hydromulch applied at a rate of 1970 kg/ha (1758 lb./ac.) provided a soil loss reduction of 19% when normalized to a straw mulch application of 2240 kg/ha (1998 lb./ac.).

Robichaud et al. [44] developed a study to evaluate the performance of wheat straw mulch and wood hydromulch when used in a post-fire condition to reduce erosion. This study utilized natural rainfall over several years to evaluate the products. Two separate tests were performed in two different locations. At the first location, the application rate of the wheat straw was 2200 kg/ha (1963 lb./ac.) and the hydromulch was 1100 kg/ha (981 lb./ac.). The soil loss reduction rates of the wheat straw mulch and the hydromulch were found to be 97% and 65%, respectively, for the first year of the study. At the second location, the application rate of the wheat straw was 4500 kg/ha to 6700 kg/ha (4015 lb./ac. to

5978 lb./ac.) and the hydromulch was 600 kg/ha (535 lb./ac.). The soil loss reduction rates of the wheat straw mulch and the hydromulch were found to be 99% and 19% for the first year, respectively.

This research aims to evaluate the effectiveness of six different ground cover treatments, normalized to a control treatment, when evaluated under simulated rainfall on laboratory scale plots. The process will include a standard and repeatable methodology that is consistently applied across the treatments under evaluation. The expected outcome is to confirm the effectiveness of the treatments.

2. Test Methods and Procedures

The validity of this research effort relies heavily on the amount of reproducible data that is collected during experiments that can be used for comparative analyses to evaluate erosion control practice and product performance and effectiveness. The test plots and rainfall simulator constructed for this research effort were replicas of Shoemaker's [45] experiments with the exception of the runoff collection device. Each test plot is 0.6 m in width by 1.2 m in length (2 ft by 4 ft) by 7.62 cm (3.5 in.) in depth. The sizes of the test plots were constructed with the purpose of testing erosion control practices with ease, speed, accuracy, and mobility throughout the experiment. The rainfall simulator was constructed using a single FullJet™ 1/2 HH—30 WSQ nozzle, with a wide angle uniform square spray area, and medium to large drop size distribution. To regulate flow rate, the inlet hose was attached to a Norgren™ R43-406-NNLA pressure regulator with 1.27 cm (1/2 in.) port sizes. To maintain a consistent pressure specific to the desired rainfall event, a pressure gauge was attached to the pressure regulator to observe and regulate operating water pressure. The simulator was suspended approximately 1.5 m (5 ft) from the building wall, and 3 m (10 ft) from the floor as shown in Figure 1, and rainfall covers approximately a 2.4 m by 2.4 m (8 ft by 8 ft) area.

Figure 1. Illustration of rainfall simulator and test plots.

Shoemaker's research efforts determined the Christiansen Uniformity Coefficient (CUC) [46] over the 2.4 m by 2.4 m (8 ft by 8 ft) spray area to range from 83% to 88% [45]; generally in the center 1.2 m by 1.2 m (4 ft by 4 ft) area.

For this study, the rainfall in 24 h for a return period of 2 years for Auburn, Alabama, was selected. The rainfall regime was designed using data available from Shoemaker [45]. The rainfall regime consisted of four separate 15-min rainfall events, each with a rainfall amount of 2.8 cm (1.1 in.) for a total rainfall amount of 11.2 cm (4.4 in.). The rainfall intensity for this regime is 11.2 cm/h (4.4 in./h). There was a 15-min period of no rainfall between two test events utilized by the researchers for data collection.

2.1. Soil Analysis

Soil for the research effort herein was provided by a local grading contractor from a construction site near the Auburn University—Erosion and Sediment Control Testing Facility (AU-ESCTF) located in Opelika, Alabama (32°33′5″ N, 85°20′28″ W, approximately 22.9 km (14.2 mi.) from Auburn, Alabama). A soil analysis was conducted by the Auburn University Soil Testing Laboratory to determine the soil composition. The experimental soil presented a "sandy clay loam" textural class according to the United States Department of Agriculture textural classification system with respective composition of 67.5%, 2.5%, and 30% of sand, silt, and clay.

After classifying the soil, a compaction test was conducted. In accordance with local standards for highway construction [47] on a typical highway embankment, slopes were compacted to 95% compaction. Given the scale of this experiment, hand tamping was selected to be used on the box plots to achieve optimum compaction. To determine the number of drops required to compact the soil, two compaction tests were completed. The first soil compaction test was to determine the optimum moisture content (OMC) or gravimetric water content of the soil. This was completed using a modified Proctor test, as specified in ASTM D1557-09, Standard Test Methods for Laboratory Compaction Characteristics of Soil Using Modified Effort [48]. The modified Proctor test enabled researchers to develop a Proctor curve representing the moisture content of the soil versus the dry unit weight of the soil, as shown in Figure 2.

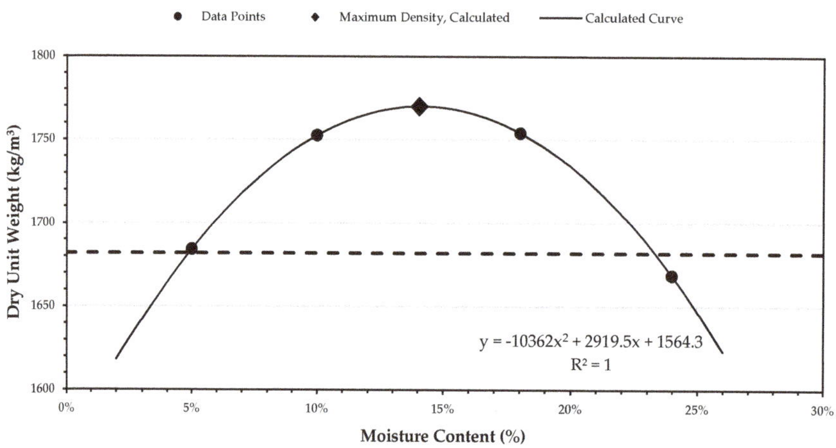

Figure 2. Proctor curve for experimental soil.

The Proctor curve shown in Figure 2 illustrates four determined moisture contents (MC) to achieve a specific dry unit weight for the tested soil. An OMC was determined to be 1762 kg/m^3 (111 lbm/ft^3 or pcf) at 14% MC by locating the maximum dry unit weight on the Proctor curve. The dotted line shown in Figure 2 represents the minimum dry unit weight of 1682 kg/m^3 (105 pcf) required to reach the specified 95% compaction rate over a MC range of 5% to 23%.

The second compaction test, also adopted from Shoemaker [45], was created to test the number of drops of the hand tamper required to achieve 95% compaction. The purpose of this compaction test was to drop the hand tamper a specified number of times upon a known volume of compacted soil to determine a corresponding unit weight. Soil with a MC of approximately 14% was loaded into the testing apparatus and a hand tamper was dropped approximately 30.5 cm (12 in.) from the soil surface in a series of 5 sets: 10, 20, 30, 50, and 60 drops. After each set of drops, the known volume of soil was weighed, and a dry unit weight was calculated, and plotted on a graph, shown in Figure 3.

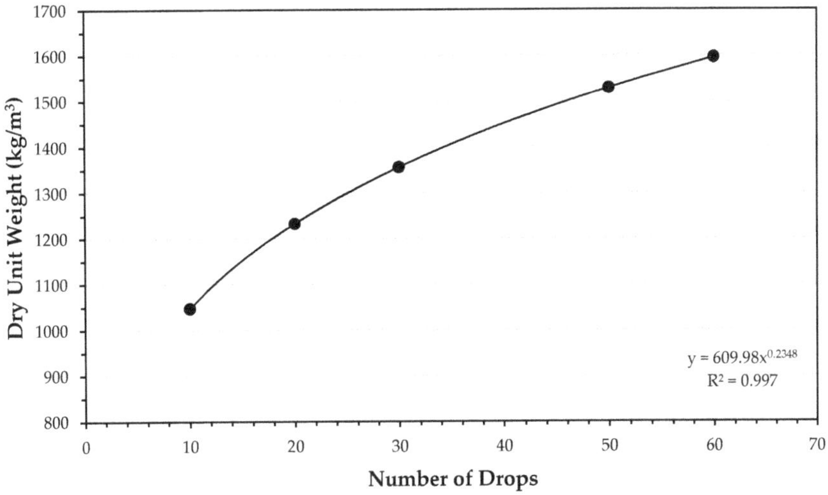

Figure 3. Number of drops with a hand tamper in relation to dry unit weight.

When compacted, soil will approach a point where it has reached maximum compaction, preventing any further compaction. A regression curve of power function was developed using the five measured points. When soil is no longer further compacted, the soil has reached maximum compaction and the dry unit weight levels off, regardless of energy applied by hand tamping. Using the power function, the specified number of drops of the hand tamper required to reach optimum compaction was calculated (Table 2).

Table 2. Calculated dry unit weight (kg/m^3) and number of required drops.

Number of Drops	Dry Unit Weight, kg/m^3 (pcf)
10	1048 (65.4)
20	1232 (76.9)
30	1355 (84.6)
40	1450 (90.5)
50	1528 (95.4)
60	1596 (99.6)
70	1655 (103.3)
80	1706 (106.5)
90	1754 (109.5)
100	1799 (112.3)

To obtain a minimum of 95% compaction, a minimum dry unit weight of 1682 kg/m^3 (105 pcf) was required, which corresponded to approximately 80 drops of the hand-tamper.

2.2. Experimental Design

Seven treatments were tested for this research effort: (1) one bare soil control; (2) conventional straw, crimped; (3) conventional straw, tackified; (4) wood fiber hydromulch; (5) straw and cotton hydromulch; (6) cotton fiber reinforced matrix hydromulch; and (7) bonded wheat fiber matrix hydromulch. Two of these treatments are classified as not having tackifiers: conventional straw, crimped and wood fiber hydromulch. The remainder of the products contain a tackifier component to the product. The bare soil treatment serves as the control, and conventional straw treatments were developed as a baseline condition for comparison of traditional mulching practices to newer hydromulch technologies currently being used in the industry. Given the application area of the rainfall simulator, two plots with the same treatment were always tested simultaneously (Figure 1) over the full experiment (four 15-min events). For each of the seven treatments tested, two separate experiments were administered; therefore, there were a total of four replicate plots for each treatment. The data for the four replicates of each treatment were averaged first before performing any further analysis.

2.3. Test Plot Preparation Prior to Condition Application

To perform this test, the soil was tested to verify the proper moisture content and then loaded into the test plots. The test plots were then compacted in a single layer of 7.62 cm (3 in.) to a density of 95% and scoured with a hand rake to a depth of 6.35 mm (1/4 in.). Once the test plots were prepared, the selected products were applied as per the manufacturer's recommended rates.

For each hydromulch product, testing was conducted using a commercially available hydroseeder (TurfMaker 380). Test boards were used to determine the number of passes required over the test plots to provide the manufacturer's specified application rates for each product. The test boards consisted of plywood with the same dimensions (0.6 m by 1.2 m (2 ft by 4 ft)) as the test plots, without the compacted soil. The applied products were scraped from the test boards and weighed to verify the application rates. The results of this testing are shown below in Table 3.

Table 3. Summary application rates for each hydromulch product.

Hydromulch Product	Manufacturer Required Dry Application Rate kg/ha (lb./ac.)	Equivalent Test Plot Required Dry Application Rate (g/plot)	Averaged Factors [1]	Minimum Number of Sprays Required
Straw and cotton HM	2241 (2000)	≈167	10.1	6
cotton FRM	3923 (3500)	≈292	9.7	7
Wood fiber HM	2241–2802 (2000–2500)	≈167–209	9.3	9
Bonded wheat FM	3362 (3000)	≈250	8.9	3

[1] Averaged factors is the product wet weight divided by the dry weight.

Once the minimum number of sprays was determined for each hydromulch product, each product was ready to be applied to test plots and tested accordingly. In order to verify application rates during the testing procedure, test boards were also sprayed in conjunction with the test plots. After the minimum number of sprays were applied to the two test boards and the two test plots, the test boards were scraped and weighed to check for application consistency to ensure manufacturer recommended rates were achieved on the test plots.

After the test plots were sprayed with the manufacturer specified application rate of the hydromulch, the test plots required time for the products to dehydrate and cure. After applying the product to the test plots, a structure was constructed, shown in Figure 4a, to hold four, 250 Watt ultraviolet-ray bulbs for the purpose of simulating natural sunlight. To ensure consistent drying, the bulbs were oriented on the structure to hang at a 3H:1V slope, which mimics the test plot setup. Lastly, the distance (approximately 45.7 cm (18 in.)) between the bulbs and the hydromulch on the test plots were measured and adjusted to ensure all bulbs were equidistant to the hydromulch surface, as illustrated in Figure 4a. The hydromulch test plots were left to dry for 48 h.

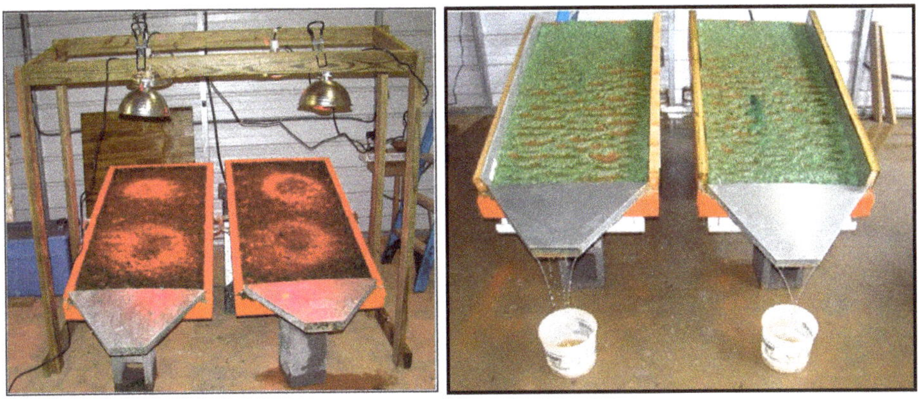

(a) (b)

Figure 4. (**a**) Drying of test plots during hydromulch testing; (**b**) collection from runoff for each test plot.

2.4. Data Collection

Collected data for this research included (1) soil loss, (2) runoff volume, and (3) turbidity. The focus was primarily on runoff generated from test plots during rainfall events. Runoff volume and mass for each 'left' and 'right' test plot (Figure 4b) was collected throughout the rain event. Instantaneous turbidity was recorded with a turbidity meter. The runoff volume and turbidity observations were recorded every minute and there were a total of 1680 observations for seven treatments on four plots for four replicates (7 × 4 × 4 × 15). The soil loss observations were recorded every 3 min (560 records = 7 × 4 × 4 × 5). Turbidity measurements were recorded from thoroughly stirred runoff collected at 1-min intervals using 4.7 L (5.0 quart) buckets.

To calculate the total soil loss, the runoff volume collected from the plots was filtered through Hayward single-length bags with one micron size pores. Once all samples were filtered, the bags were placed in an oven at 71.1 °C (160 °F) and dried for 24 h. After drying, the bags were compared to the weight of the empty bags recorded prior to filtering to determine the amount of eroded soil from each test plot contained within each bag.

2.5. Statistical Analyses

The Tukey–Kramer method, a single-step multiple comparison procedure and statistical test, was used to analyze the recorded data and establish statistical significance between treatments [45].

3. Results and Discussion

3.1. Turbidity Variations

Using the previously outlined procedures, turbidity measurements were recorded for each series of tests from a thoroughly stirred bucket of runoff collected at 1-min intervals. A summary of the collected results is provided below in Table 4. Average turbidity of all four replicate plots for each minute and each treatment is presented in Figure 5 for four 15-min events for the bare soil (Control) and six erosion control treatments. When compared to the bare soil treatment, labeled 'Control', turbidity was reduced by at least a factor of 6 for all treatments by the end of the 60 min test ('Event 4').

Table 4. Average turbidity, standard deviation, and percent reduction of each treatment with respect to the control of four 15-min events for surface runoff.

Treatment	Average Turbidity (NTU) [1]	Standard Deviation (NTU)	Percent Reduction
Control	6060	638	-
Straw and cotton HM	501 [a]	150	92%
Straw, crimped	1240 [a,b]	468	80%
Cotton FRM	277 [a,c]	71	95%
Wood fiber HM	930 [a,b,d]	285	85%
Bonded wheat FM	59 [a,b,c,e]	10	99%
Straw, tackified	148 [a,b,c,e,f]	35	98%

[1] Letters following the value show whether it is significantly different ($p < 0.05$) to the referenced treatment: [a] represents significantly different to the control; [b] represents significantly different to straw and cotton HM; [c] represents significantly different to straw, crimped; [d] represents significantly different to cotton FRM; [e] represents significantly different to wood fiber HM; [f] represents significantly different to bonded wheat FM.

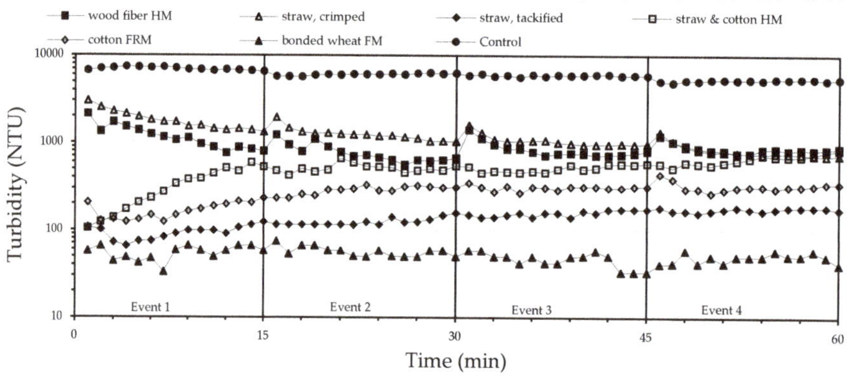

Figure 5. Average turbidity of surface runoff vs. time. Average turbidity for each minute was calculated for all four replicate plots for each treatment.

As shown in Figure 5, each hydromulch with the exception of the wood fiber HM and the straw and cotton HM were capable of reducing turbidity levels to under 500 NTUs. Two observations can be made from Figure 5: (1) the treatments without a polymer-enhanced tackifier (e.g., conventional straw, crimped, and wood fiber HM) had higher turbidity values during 'Event 1' and 'Event 2', whereas the turbidity decreased slightly during the last two rainfall events in comparison to treatments with a tackifier; (2) the treatments with tackifiers started with very low turbidity values and steadily increased over the four, 15 min rainfall events. The bonded wheat FRM was the only product to maintain a steady turbidity of about 60 NTUs throughout the four rainfall events. The improved performance of the treatments containing a tackifier in comparison to the treatments without a tackifier is likely due to the bonding of the tackifier with the soil particles, which in turn creates a more hydrophobic environment [38].

Table 4 shows average turbidity measurements, standard deviation of the average turbidity, and a percent reduction, normalized for the control condition. As shown, the bonded wheat FRM is the most effective treatment in reducing average turbidity of nearly 99%, followed by straw, tackified, cotton FRM, straw and cotton HM, wood fiber HM, and straw, crimped with percent reductions of 98%, 95%, 92%, 85%, and 80% respectively. A statistical analysis was conducted and the values for average turbidity were compared to determine if the results were statistically significantly different. The results are denoted by different letters as shown in Table 4: [a] represents significantly different to the control; [b] represents significantly different to straw and cotton HM; [c] represents significantly different to straw, crimped; [d] represents significantly different to cotton FRM; [e] represents significantly different to wood fiber HM; [f] represents significantly different to bonded wheat FM. Shoemaker [35] also computed and reported the lower and upper bounds of confidence intervals for all comparisons.

Hydromulches typically include tackifying or bonding agents to bond the mulch particles to the soil surface. Once the hydromulch dries on the soil surface, a crusted, rough surface is formed which is typically a more hydrophobic environment. The crusted surface is designed to absorb the rainfall and serve as a filtration system to capture soil particles suspended in the stormwater runoff. When the tackifier or bonding agents have been washed away or begin to degrade due to stormwater runoff, the turbidity observed began to increase slightly as shown in Figure 5 above for the straw, tackified, cotton FRM, and straw and cotton HM. However, products with stronger tackifying agents such as bonded wheat FM take longer to deteriorate.

The treatments without a tackifier, straw, crimped and wood fiber HM, rely primarily on the mulch material by itself to minimize erosion from the plots. From a soil erosion perspective, these treatments are functioning as a protective layer to minimize the splash erosion created by the rainfall. Splash erosion has been found to be the initial cause of erosion [49]. An observation was made from Figure 5 during the first two rainfall events, which was that the treatments that do not have a tackifying agent applied experienced a higher rate of erosion due to the absence of a tackifying agent to bond the soil particles to the treatment. This initial large concentration of soil in the runoff at the beginning of a rainfall event is due to the splash erosion caused by the raindrops impacting the soil surface. The treatments which contain a tackifying agent lessen this initial erosion by bonding the soil particles with the other material. On the other hand, the products without a tackifying agent lessen the amount of splash erosion by providing a surface cover over the soil particles when compared to the bare soil treatment.

A statistical analysis was completed to confirm observed differences between the control and treatments for turbidity measurements of stormwater surface runoff. ANOVA tables were created using Tukey–Kramer comparison tests to determine statistical significance between individual pairs of groups, as illustrated in Table 4. As observed, this table demonstrated that the average turbidities had statistically significant differences between the control and all treatments. All treatments showed significant differences between them in the average turbidity except for straw and cotton HM and cotton FRM. Additionally, no significant statistical difference was observed between cotton FRM and bonded wheat FM, cotton FRM and straw, tackified, and bonded wheat FM and straw, tackified. All other treatment comparisons proved to show a statistically significant difference as shown in Table 4.

3.2. Soil Loss

Samples used to calculate soil loss were collected from simulated rainfall runoff every 3 min for all experiments conducted. Based on the data collected, it was observed that all treatments had significantly smaller levels of soil loss when compared to the bare soil (control). The control condition and the treatments without a tackifying agent (i.e., straw, crimped, and wood fiber HM) experienced an initial surge of soil loss due to the breakage of soil aggregates by the impact of raindrops, with the consequent dispersion of fine particles (splash erosion). However, the treatments with tackifiers did not have this surge; a steady increase in soil loss over time for each rainfall event was observed for these treatments. As shown in Figure 6, the most effective treatment in reducing soil loss was bonded wheat FM. After the first rainfall event, it was observed that soil loss measurements remained consistent for the remainder of the experiment. The summarized data is provided in Figure 6 below.

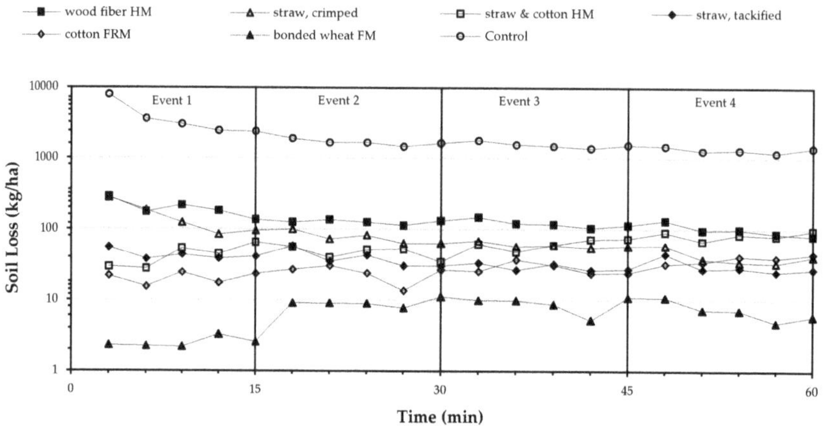

Figure 6. Three-minute soil loss vs. time for all treatments as compared to the control.

The control recorded more soil loss than all of the treatments in the first rainfall event by a factor of 17. The most consistent and effective erosion control treatment was bonded wheat FM, maintaining an average soil loss of approximately 11.2 kg/ha (10 lb./ac.) over the entire experiment. Wood fiber HM was observed to produce the largest consistent amount of eroded soil, starting at approximately 1008 kg/ha (900 lb./ac.), and decreasing to approximately 504 kg/ha (450 lb./ac.) by the last rainfall event. Straw and cotton HM showed initial signs of strength in controlling erosion with 224 kg/ha (200 lb./ac.) of cumulative eroded soil, however steadily increased to almost 448 kg/ha (400 lb./ac.) by 'Event 4', nearly doubling its initial amount. It was also observed that straw, crimped began with approximately the same amount of cumulative soil loss as wood fiber HM; however after the first two rainfall events, steadily decreased to nearly 224 kg/ha (200 lb./ac.), which are soil loss levels similar to that of straw, tackified, and cotton FRM. The cotton FRM averaged 112 kg/ha (100 lb./ac.) over the entire experiment. This data is shown in Figure 7 below.

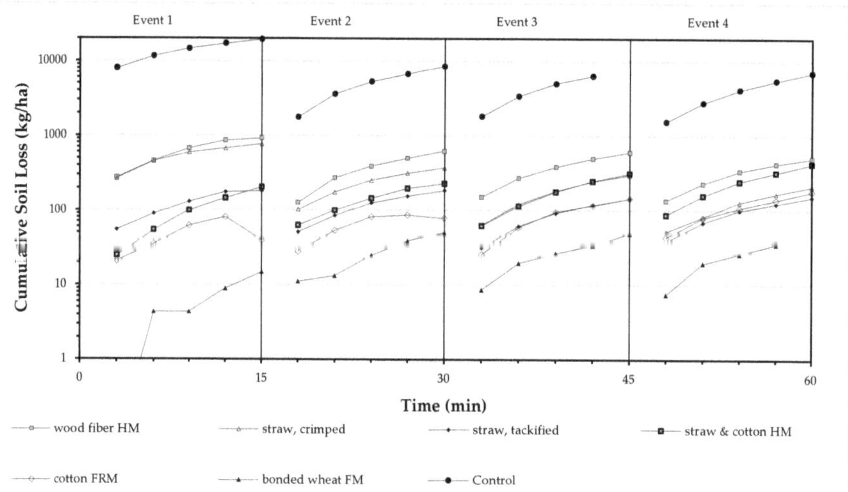

Figure 7. Cumulative soil loss vs. time for six treatments as compared to the control.

Table 5 presents specific values of average soil loss, standard deviation, and percent reduction for each treatment during each rainfall event. The straw, crimped treatment, when normalized to the control, reduced erosion during the first rainfall event by nearly 96% and increased to approximately 98.9% by the fourth rainfall event. Similarly, straw, tackified, and wood fiber HM increased in percent reduction from 'Event 1' to 'Event 4' by 98.9% to 99.2% and 94.9% to 97.4%, respectively. The hydromulches with tackifying agents reacted in a dissimilar way when normalized to the control. Over the rainfall events, percent reductions decreased from 98.9%% to 97.8%, 99.5 to 99.1%, and 99.9% to 99.7% for straw and cotton HM, cotton FRM, and bonded wheat FM, respectively. It was observed that this reduction was due to the degradation of the tackifying bonds between the soil and the mulch; contrarily, the increased performance of the non-tackified treatments was observed to be due to the 'flush effect' of the scoured surface in the first events, exposing the less erodible, compacted, underlying soil.

Table 5. Average soil loss over each 15-min rainfall event due to surface runoff.

Condition	Soil Loss [1] (kg/ha)	Standard Deviation [2] (kg/ha)	Percent Reduction [3] (%)
		1st 15-min rainfall event	
Control	3889	3000	-
Straw, crimped	155.4	108.0	96.0
Straw, tackified	42.8	39.6	98.9
Wood fiber HM	198.9	158.4	94.9
Straw and cotton HM	42.8	50.1	98.9
Cotton FRM	20.7	12.9	99.5
Bonded wheat FM	3.81	2.9	99.9
		2nd 15-min rainfall event	
Control	1694	212.7	-
Straw, crimped	77.6	24.3	98.0
Straw, tackified	38.3	38.7	99.0
Wood fiber HM	127.1	84.7	96.7
Straw and cotton HM	46.9	39.5	98.8
Cotton FRM	23.8	16.9	99.4
Bonded wheat FM	11.2	3.70	99.7
		3rd 15-min rainfall event	
Control	1602	264.9	-
Straw, crimped	62.0	135.6	98.4
Straw, tackified	30.2	26.7	99.2
Wood fiber HM	121.4	84.4	96.9
Straw and cotton HM	64.3	53.1	98.3
Cotton FRM	29.7	18.3	99.2
Bonded wheat FM	10.5	5.8	99.7
		4th 15-min rainfall event	
Control	1377	217.5	-
Straw, crimped	44.1	13.8	98.9
Straw, tackified	31.7	28.4	99.2
Wood fiber HM	101.8	78.0	97.4
Straw and cotton HM	84.3	56.3	97.8
Cotton FRM	36.7	24.5	99.1
Bonded wheat FM	10.3	3.7	99.7

[1] Average of 3-min soil loss (Figure 6) for each 15-min rainfall event. [2] Standard deviation for average soil loss over an event. [3] Denotes values normalized by control condition.

Continuing the statistical analysis used throughout this research effort, ANOVA procedures with Tukey–Kramer multiple comparison tests were used for the recorded amounts of soil loss. Table 6 illustrates statistically significant and insignificant results of average soil loss throughout the experiments. The statistical analysis compared all treatments to the control and each other. The control proved to be statistically different to all treatments; therefore, each treatment had a significant effect in reducing soil loss when compared the bare soil. No significant differences were found between the comparison to the other treatments. Therefore, it can be concluded from Table 6 that statistically, each treatment is capable of significantly reducing and controlling erosion on 3H:1V, compacted fill slopes.

Table 6. Cumulative soil loss for four, 15-min rainfall events (A) and calculated soil loss ratio per treatment.

Treatment	Cumulative Soil Loss (A) (grams/plot) [2]	Cumulative Soil Loss (A) (g/m^2)	* Calculated Soil Loss Ratio [1]
Straw, crimped	126 [a]	169.5	0.040
Straw, tackified	53 [a]	71.3	0.017
Wood fiber HM	204 [a]	274.4	0.064
Straw and cotton HM	89 [a]	119.7	0.028
Cotton FRM	41 [a]	55.2	0.013
Bonded wheat FM	13 [a]	17.5	0.004

[1] Soil loss ratio normalized to a bare soil value of 4281 g/m^2. * Soil loss ratio calculation: SLR = A/Control [1]. [2] The letter [a] following the values show that they are significantly different ($p < 0.05$) to the control.

3.3. Cover-Factor

Several studies [39,41,43,50,51] used a 'cover-factor' to report erosion control performance. The cover factor is a parameter in the Revised Universal Soil Loss Equation (RUSLE) to represent a comparison of soil loss occurring with the treatment in place to that which occurs in the bare, unprotected condition [51]. The RUSLE allows researchers to calculate cover-factors for treatments without testing a bare soil using several different parameters based upon soil type, slope, and rain regimes; ECTC [50] and Clopper et al. [51] used the RUSLE to calculate cover-factors. However, in this study, the treatment results were compared to the results of the bare soil control test. This comparison is defined as the "Soil-Loss Ratio". Table 6 summarizes the soil loss ratio calculated in this research effort. According to calculated soil loss ratios of 0.004, 0.013, 0.017, 0.028 0.040, and 0.064 in Table 6, the hydromulches can be ranked from most to least effective erosion control practices accordingly: (1) bonded wheat FM, (2) cotton FRM, (3) straw, tackified, (4) straw and cotton HM, (5) straw, crimped, and (6) wood fiber HM.

4. Conclusions

Twenty eight experiments were conducted to examine the ESC effectiveness of seven treatments: (1) control (bare soil), (2) conventional straw, crimped, (3) conventional straw, tackified, (4) wood fiber HM, (5) straw and cotton HM, (6) cotton FRM, and (7) bonded wheat FM. Performance was evaluated using data collection from experiments, which included surface runoff volume, soil loss, and turbidity.

Turbidity measurements were recorded from samples that were collected every minute of each of the four, 15 min rainfall events. The order of the six treatments ranked from most to least effective according to an averaged percent reduction when normalized by the bare soil condition were (1) bonded wheat FM (99% reduction), (2) straw, tackified (98% reduction), (3) cotton FRM (95% reduction), (4) straw and cotton HM (92% reduction), (5) wood fiber HM (85% reduction), and (6) straw, crimped (80% reduction). The erosion control practices without tackifiers (conventional straw, crimped, and wood fiber HM) experienced a significant increase in the amount of eroded sediment during the first two rainfall events, likely caused by the lack of the bonding between soil particles; however, each treatment steadily improved sediment control over time. Contrarily, the surface cover practices with tackifying agents provided excellent initial sediment control due to the bonds between soil particles and the mulching materials, but over the four rainfall events, the chemical bonds began to deteriorate, showing a steady decrease in performance.

Approximately, 100%, 99%, 98%, 97%, 96%, and 94% soil loss reduction for bonded wheat FM, cotton FRM, straw-tackified, straw and cotton HM, straw-crimped, and wood fiber HM were observed, respectively. Cumulative soil losses were also used in this research to calculate the soil loss ratios (SLR) between treated and untreated conditions; calculated ratios mimicked percent reduction performances, ranging in value from 0.004 for bonded wheat FM to 0.064 for wood fiber HM.

Literature reviewed and results from this research effort suggest that conventional straw crimped or tackified as well as hydromulches are very effective erosion control measures, when applied at the proper application rates.

Recommended Future Research

Results and conclusions presented in this study show that conventional straw (crimped or tackified) and hydromulches are effective means of reducing erosion and sedimentation caused by sediment laden runoff. However, using soil specific polyacrylamide would likely result in greater erosion control than the non-soil specific polymers that make up many of the tackifying agents used on straw and that are a part of many hydromulch products. Therefore, further research should be conducted to examine how the addition of polyacrylamide to these practices could potentially improve in-field performance.

Additionally, the conclusions derived in this study are based on laboratory-scale test plots. It would be beneficial if the performance of these treatments were tested at field-scale conditions to validate laboratory-scale results provided in this research. Laboratory-scale experiments allow researchers to test the performance of erosion control practices at a faster rate than most field-scale experiments; therefore, if field-scale test results are similar to laboratory-scale test results, a larger quantity of products could be effectively evaluated in a shorter period of time using laboratory-scale tests.

Author Contributions: M.D.R. prepared the manuscript draft. W.T.W. designed the rainfall simulator, conducted the simulations and analysis of the results, and prepared the manuscript draft, W.C.Z. supervised the design and simulations and revised the manuscript, X.F. and W.N.D. provided input on the writing, data analysis, and revised the manuscript. All authors made contributions to the study and writing of the manuscript. All authors have read and agreed to the published version of the manuscript.

Funding: This research was funded by the Auburn University Highway Research Center.

Acknowledgments: The authors would like to thank the Highway Research Center at Auburn University for its sponsorship of this research. The authors gratefully acknowledge this financial support. The findings, opinions, and conclusions presented are those of the authors and do not necessarily reflect the view of the sponsors.

Conflicts of Interest: The authors declare no conflict of interest.

References

1. Zech, W.C.; Halverson, J.L.; Clement, T.P. Development of a Silt Fence Tieback Design Methodology for Highway Construction Installations. *J. Transp. Res. Board Transp. Res. Rec.* **2007**, *2011*, 21–28. [CrossRef]
2. United States Environmental Protection Agency (US EPA) Clean Water Act Section 502: General Definitions. Available online: https://www.epa.gov/cwa-404/clean-water-act-section-502-general-definitions (accessed on 19 December 2019).
3. United States Enviromental Protection Agency (US EPA). *Innovative Uses of Compost: Erosion Control., Turf Remediation and Landscaping*; United States Enviromental Protection Agency: Washington, DC, USA, 1997.
4. United States Environmental Protection Agency (US EPA). *Stormwater Phase II Final Rule: Construction Site Runoff Control Minimum Control Measure*; United States Enviromental Protection Agency: Washington, DC, USA, 2005.
5. Scholl, B.N.; Holt, G.; Thornton, C.; Duke, S. Hydromulch blends using agricultural byproducts: Performance implications of cotton quantity. *J. Cotton Sci.* **2013**, *17*, 302–308.
6. Megahan, W.F.; Wilson, M.; Monsen, S.B. Sediment production from granitic cutslopes on forest roads in Idaho, USA. *Earth Surf. Processes Landf.* **2001**, *26*, 153–163. [CrossRef]
7. Risse, M.; Faucette, B. *Compost Utilization for Erosion Control*; University of Georgia: Athens, GA, USA, 2001.
8. United States Environmental Protection Agency (US EPA). *Stormwater Phase II Final Rule: An Overview*; United States Environmental Protection Agency: Washington, DC, USA, 2005.

9. United States Environmental Protection Agency (US EPA). *2000 National Water Quality Inventory*; United States Environmental Protection Agency: Washington, DC, USA, 2000.
10. Novotny, V. *Water Quality: Diffuse Pollution and Watershed Management*; John Wiley & Sons, Inc.: New York, NY, USA, 2003.
11. Brady, N.C.; Weil, R.R. *The Nature and Properties of Soils*, 14th ed.; Prentice Hall: Upper Saddle River, NJ, USA, 2014.
12. District, M.H.F. Chapter 7 Construction BMPs. Available online: https://udfcd.org/wp-content/uploads/uploads/vol3criteriamanual/Chapter7ConstructionBMPs.pdf (accessed on 24 November 2019).
13. Scholl, B.N.; Holt, G.; Thornton, C. Screening study of select cotton-based hydromulch blends produced using the Cross-Linked biofiber process. *J. Cotton Sci.* **2012**, *16*, 249–254.
14. Tyner, J.S.; Yoder, D.C.; Chomicki, B.J.; Tyagi, A. A Review of Construction Site Best Management Practices for Erosion Control. *Trans. ASABE* **2011**, *54*, 441–450. [CrossRef]
15. Alabama Soil and Water Conservation Committee (ASWCC). *Alabama Handbook for Erosion Control., Sediment Control, and Stormwater Management on Construction Site and Urban Areas: Volume 2*; Alabama Soil and Water Conservation Committee: Montgomery, AL, USA, 2009.
16. Box, J.E.J.; Bruce, R.R. The effect of surface cover on infiltration and soil erosion. In *Soil Erosion, Conservation and Rehabilitation*; Aggasi, M., Ed.; Dekker: New York, NY, USA, 1996; pp. 107–123.
17. Bruce, R.R.; Langdale, G.W.; West, L.T.; Miller, W.P. Surface Soil Degradation and Soil Productivity Restoration and Maintenance. *Soil Sci. Soc. Am. J.* **1995**, *59*, 654–660. [CrossRef]
18. Sutherland, R.A. Rolled erosion control systems for hillslope surface protection: A critical review, synthesis and analysis of available data. I. Background and formative years. *Land Degrad. Dev.* **1998**, *9*, 465–486. [CrossRef]
19. Sutherland, R.A.; Ziegler, A.D. Hillslope runoff and erosion as affected by rolled erosion control systems: A field study. *Hydrol. Processes Int. J.* **2006**, *20*, 2839–2855. [CrossRef]
20. Rickson, R.J. Simulated vegetation and geotextiles. In *Slope Stabilization and Erosion Control: A Bioengineering Approach*; Morgan, R., Rickson, R., Eds.; Spon: London, UK, 1995; pp. 95–131.
21. Turgeon, A.J. Turfgrass Management. In *Turfgrass Management*; Prentice Hall: Upper Saddle River, NJ, USA, 2002.
22. Singer, M.T.; Matsuda, Y.; Blackard, J. Effect of Mulch Rate on Soil Loss by Raindrop Splash. *Soil Sci. Soc. Am. J.* **1981**, *45*, 107–110. [CrossRef]
23. Prats, S.A.; Malvar, M.C.; Vieira, D.C.S.; MacDonald, L.; Keizer, J.J. Effectiveness of Hydromulching to Reduce Runoff and Erosion in a Recently Burnt Pine Plantation in Central Portugal. *Land Degrad. Dev.* **2016**, *27*, 1319–1333. [CrossRef]
24. Keizer, J.J.; Silva, F.C.; Vieira, D.C.S.; González-Pelayo, O.; Campos, I.; Vieira, A.M.D.; Valente, S.; Prats, S.A. The effectiveness of two contrasting mulch application rates to reduce post-fire erosion in a Portuguese eucalypt plantation. *Catena* **2018**, *169*, 21–30. [CrossRef]
25. Babcock, D.; McLauglin, R.A. Erosion control effectiveness of straw, hydromulch, and polyacrylamide in a rainfall simulator. *J. Soil Water Conserv.* **2013**, *68*, 221–227. [CrossRef]
26. Mannering, J.V.; Meyer, L.D. The Effects of Various Rates of Surface Mulch on Infiltration and Erosion. *Soil Sci. Soc. Am. J.* **1967**, *29*, 84–85. [CrossRef]
27. McLauglin, R.A.; Brown, T.T. Evaluation of erosion control products with and without added polyacrylamide. *J. Am. Water Resour. Assoc.* **2006**, *42*, 675–684. [CrossRef]
28. Meyer, L.D.; Wischmeier, W.H.; Foster, G.R. Mulch Rates Required for Erosion Control on Steep Slopes. *Soil Sci. Soc. Am. J.* **1970**, *34*, 928–930. [CrossRef]
29. Babcock, D.; McLauglin, R.A. Mulch Options for Erosion Control on Construction Sites. Available online: https://content.ces.ncsu.edu/mulch-options-for-erosion-control-on-construction-sites (accessed on 24 November 2019).
30. Lee, G.; McLaughlin, R.A.; Whitely, K.D.; Brown, V.K. Evaluation of seven mulch treatments for erosion control and vegetation establishment on steep slopes. *J. Soil Water Conserv.* **2018**, *73*, 434–442. [CrossRef]
31. Babcock, D.L.; McLauglin, R.A. Runoff water quality and vegetative establishment for groundcovers on steep slopes. *J. Soil Water Conserv.* **2011**, *66*, 132–141. [CrossRef]
32. McCullough, S.A.; Endress, B.A. Do postfire mulching treatments affect plant community recovery in California coastal sage scrub lands? *Environ. Manag.* **2012**, *49*, 142–150. [CrossRef]

33. Eck, B.; Barrett, M.; McFarland, A.; Hauck, L. Hydrologic and Water Quality Aspects of Using a Compost/Mulch Blend for Erosion Control. *J. Irrig. Drain. Eng.* **2010**, *136*, 646–655. [CrossRef]
34. Morgan, R.; Rickson, J. Soil erosion control: Importance of geomorphological information. In *Geomorphology in Environmental Planning*; Hooke, J.M., Ed.; Wiley: Chichester, UK, 1988; pp. 51–60.
35. National Cooperative Highway Research Program (NCHRP). *Erosion Control During Highway Construction—Research Report*; National Cooperative Highway Research Program: Washington, DC, USA, 1980.
36. Weggel, J.R.; Rustom, R. Soil erosion by rainfall and runoff—State of the art. *Geotext. Geomembr.* **1992**, *11*, 551–572. [CrossRef]
37. Erosion Control Technology Council (ECTC) Erosion Control Product Categories. Available online: https://www.ectc.org/product-selection-by-category (accessed on 24 November 2019).
38. Vaughn, S.F.; Kenar, J.A.; Felker, F.C.; Berhow, M.A.; Cermak, S.C.; Evangelista, R.L.; Fanta, G.F.; Behle, R.W.; Lee, E. Evaluation of alternatives to guar gum as tackifiers for hydromulch and as clumping agents for biodegradable cat litter. *Ind. Crops Prod.* **2013**, *43*, 798–801. [CrossRef]
39. Holt, G.A.; Buser, M.; Harmel, D.; Potter, K.; Pelletier, M. Comparison of Cotton-based Hydro-mulches and Conventional Wood and Paper Hydro-mulches-Study 1. *J. Cotton Sci.* **2005**, *9*, 121–127.
40. Holt, G.A.; Laird, J.W. COBY Products and A Process for Their Manufacture 2002. U.S. Patent 6,383,548, 7 May 2002.
41. Landloch. *Studies of Hydromulch Effectiveness*; National Center for Engineering in Agriculture: Toowoomba, Australia, 2002.
42. Benik, S.R.; Wilson, B.N.; Biesboer, D.D.; Hansen, B.; Stenlund, D. Performance of erosion control products on a highway embankment. *Trans. ASAE* **2003**, *46*, 1113–1119. [CrossRef]
43. Buxton, H.; Caruccio, F.T. *Evaluation of Selective Erosion Control. Techniques*; United States Environmental Protection Agency: Washington, DC, USA, 1979.
44. Robichaud, P.R.; Lewis, S.A.; Wagenbrenner, J.W.; Ashmun, L.E.; Brown, R.E. Post-fire mulching for runoff and erosion mitigation Part I: Effectiveness at reducing hillslope erosion rates. *Catena* **2013**, *105*, 75–92. [CrossRef]
45. Shoemaker, A.L. Evaluation of Anionic Polyacrylamide as an Erosion Control Measure Using Intermediate-Scale Experimental Procedures. Master's Thesis, Auburn University, Auburn, AL, USA, 2009.
46. ASTM International. *Standard Test Method for Determination of Rolled Erosion Control Product Performance in Protecting Hillslopes from Rainfall-Induced Erosion*; ASTM International: West Conshohocken, PA, USA, 2019.
47. ALDOT. *Standard Specifications for Highway Construction*; Alabama Department of Transportation: Montgomery, AL, USA, 2018.
48. ASTM International. *ASTM D1557-09, Standard Test Methods for Laboratory Compaction Characteristics of Soil Using Modified Effort (56,000 ft-lbf/ft3 (2700 kN-m/m3))*; ASTM International: West Conshohocken, PA, USA, 2009.
49. Angulo-Martínez, M.; Beguería, S.; Navas, A.; Machín, J. Splash erosion under natural rainfall on three soil types in NE Spain. *Geomorphology* **2012**, *175–176*, 38–44. [CrossRef]
50. Erosion Control Technology Council (ECTC). Comparison of Erosion Control Technologies: Blown Straw vs. Erosion Control Blanket, Quantification of Performance and Value. Available online: https://www.ectc.org/assets/docs/Whitepapers/ectc_june2015_blown%20straw%20vs%20recp.pdf (accessed on 11 February 2020).
51. Clopper, P.E.; Vielleux, M.; Johnson, T. *Quantifying the Performance of Hillslope Erosion Control Best Management Practices*; World Water and Environmental Resources Congress: Orlando, FL, USA, 2001.

water

MDPI

Article

Spatial and Temporal Patterns of Rainfall Erosivity in the Tibetan Plateau

Zhijia Gu [1,2], Detai Feng [2], Xingwu Duan [2,3,*], Kuifang Gong [4], Yawen Li [2] and Tianyu Yue [5]

1 School of Geographic Sciences, Xinyang Normal University, Xinyang 464000, China; guzhijia@hotmail.com
2 Institute of International Rivers and Eco-security, Yunnan University, Kunming 650091, China; feng_detai@126.com (D.F.); yawenlynu@163.com (Y.L.)
3 Yunnan Key Laboratory of International Rivers and Trans-boundary Eco-security, Yunnan University, Kunming 650091, China
4 Soil and Water Conservation Bureau of Tibet Autonomous District, Lhasa 850000, China; comgu.ya.nan@163.com
5 Faculty of Geographical Science, Beijing Normal University, Beijing 100875, China; tyyue@mail.bnu.edu.cn
* Correspondence: xwduan@ynu.edu.cn

Received: 19 December 2019; Accepted: 8 January 2020; Published: 10 January 2020

check for updates

Abstract: The Tibetan Plateau is influenced by global climate change which results in frequent melting of glaciers and snow, and in heavy rainfalls. These conditions may increase the risk of soil erosion, but prediction is not feasible due to scarcity of rainfall data in the high altitudes of the region. In this study, daily precipitation data from 1 January 1981 to 31 December 2015 were selected for 38 meteorological stations in the Tibetan Plateau, and annual and seasonal rainfall erosivity were calculated for each station. Additionally, we used the Mann–Kendall trend test, Sen's slope, trend coefficient, and climate tendency rate indicators to detect the temporal variation trend of rainfall erosivity. The results showed that the spatial distribution of rainfall erosivity in the Tibetan Plateau exhibited a significant decreasing trend from southeast to northwest. The average annual rainfall erosivity is 714 MJ·mm·ha^{-1}·h^{-1}, and varies from 61 to 1776 MJ·mm·ha^{-1}·h^{-1}. Rainfall erosivity was mainly concentrated in summer and autumn, accounting for 67.5% and 18.5%, respectively. In addition, annual, spring, and summer rainfall erosivity were increasing, with spring rainfall erosivity highly significant. Temporal and spatial patterns of rainfall erosivity indicated that the risk of soil erosion was relatively high in the Hengduan mountains in the eastern Tibetan Plateau, as well as in the Yarlung Zangbo River Valley and its vicinity.

Keywords: rainfall erosivity; soil erosion; spatial and temporal pattern; Mann–Kendall test; Tibetan Plateau

1. Introduction

Soil erosion is a global environmental problem, which leads to land degradation, siltation of reservoirs, and eutrophication of water bodies, among others [1–3]. Formation mechanisms and succession processes of soil erosion are affected by rainfall erosivity, which is highly correlated with the product (EI) of the total storm energy (E) and the maximum 30 min intensity (I_{30}), both derived from data by Wischmeier and Smith (1958) [4] and Wischmeier (1959) [5]. The concept of rainfall erosivity developed further by Hudson (1971) [6] and Wischmeier and Smith (1978) [7] describes erosivity as the average of annual summations of storm EI_{30}. Rainfall erosivity is the basic factor in the Universal Soil Loss Equation [7] and the Revised Universal Soil Loss Equation [8]. Additionally, many empirical soil erosion predictions also use rainfall erosivity [9–12]. Although the EI_{30} index has been accepted worldwide, its calculation requires hyetograph data for a storm. Therefore, the use of the EI_{30} index has been limited by a lack of high temporal resolution rainfall data.

Many statistical models have been established based on annual rainfall [13–17], monthly rainfall [18–20], daily rainfall [21–25], and even hourly rainfall [26,27]. Hourly rainfall data cannot be obtained from many of the national meteorological observation stations, which limits their application. By contrast, daily rainfall data can guarantee the accuracy of rainfall erosivity calculations due to the level of detail on rainfall variability [21]. Daily rainfall data is widely used in the calculation of rainfall erosivity [21–25].

The spatial heterogeneity of soil erosion is very high and has multiple scale characteristics. It depends on the diversity and complexity of factors affecting soil erosion. Rainfall erosivity is one of the basic factors responsible for soil erosion [28]. However, it is not distributed uniformly throughout the year, and at regional scale, knowledge of seasonal and even semi-monthly distribution of rainfall erosivity is critical to the accuracy of soil erosion calculations [8,11]. Therefore, the spatial and temporal distribution of rainfall erosivity concerns various countries and regions [12,15,29–32]. Spatial variability in rainfall erosivity in China is relatively large, with rainfall erosivity decreasing from the southeast to the northwest [33,34]. Therefore, compared with the northwest region, the risk of erosion in the southeastern region was significantly greater, and the area is also receiving the attention of scientists.

The global climate system is undergoing a change characterized by significant warming [35]. The Tibetan Plateau is sensitive to climate change and ecologically fragile [36]. Climate in the Tibetan Plateau has changed dramatically in the early 21st century, with the continuing severe warming and increasing precipitation [37]; changes include increasing snowmelt and more frequent heavy precipitation events. Vegetation changes in the Tibetan Plateau and its response to climate change [38], precipitation and climate change [39], runoff and soil erosion [40–42], as well as watershed non-point source pollution in the main farming area [43], have attracted the attention of researchers. Protecting the grasslands of the Tibetan Plateau is of great importance in limiting global climate change [44]. However, grassland degradation phenomena have been found due to the unreasonable use of grassland by human beings and changing global climate [45]. Besides, aeolian desertification is impeding sustainable socio-economic development [46]. Furthermore, soil erosion in this area is relatively weak compared to the eastern part of China, and recovery from soil erosion is lengthy and difficult once it occurs.

The service span of water conservancy facilities and flood control capabilities were reduced by severe soil erosion that led to the siltation of channel beds and thus a reduction in reservoir capacity in the Tibetan Plateau [47]. The Tibetan Plateau, especially the gorge area in the eastern part, has a large elevation difference where severe soil erosion restricts the development of agriculture and animal husbandry [42]. Sparse distribution of meteorological stations in this region prevented data collection on rain characteristics; however, surface runoff generated by snowmelt can also cause soil erosion. Although researchers have done some studies on rainfall erosivity in the Tibetan Plateau, it is far from enough [48–51]. Dynamic monitoring of soil erosion in China still requires detailed data on rainfall erosivity in the area.

To address this, daily precipitation data were used from 1 January 1981 to 31 December 2015 for a total of 38 stations in Tibet, and the annual and seasonal rainfall erosivity was calculated. The objectives of the study were to (a) identify long-term trends in rainfall erosivity, and (b) map the spatiotemporal patterns in rainfall erosivity. The results of this paper are intended to optimize the quantitative prediction of soil erosion and soil and water conservation planning services in the Tibetan Plateau.

2. Materials and Methods

2.1. Study Area

The Tibetan Plateau is part of and located in the southwestern part of the Qinghai–Tibet Plateau. It has a total area of about 1.2 million km^2 and an average elevation of 4000 m, which gave it its designation as the "roof of the world". Most of the Tibetan Plateau is arid and semi-arid. Precipitation has an uneven distribution and large spatial differentiation. The southeast area has annual precipitation of 600–800 mm, while the western area suffers from drought with annual precipitation of less than 200 mm. Precipitation from May through October accounts for more than 80% of the annual total. The spatial distribution of precipitation indicates a decreasing trend from the southeast to the northwest [52]. Mean annual temperature varies between −3.0 and 11.8 °C, and daily temperature fluctuations are often as much as about 15 °C [53]. The region experiences soil erosion, desertification, and mountain hazards. Grasslands make up the largest proportion of land cover, accounting for 56.72% of the total land area in the region. The proportion of cultivated land is 0.42% and it is mainly found in southern Tibet, with a small amount distributed in the east and southeast. Due to the high altitude in the northwestern region, the snow cover and frozen soil make this area unsuitable for human habitation, and the level of human activities is thus low. Thirty-eight meteorological stations used in this study are primarily found in the eastern and southern parts of the Tibetan Plateau, with very few in the northwest (Figure 1).

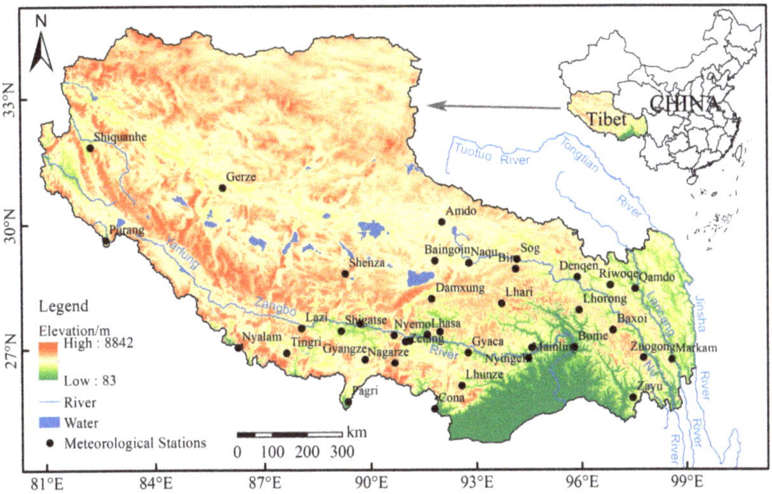

Figure 1. Study area and distribution of meteorological stations.

2.2. Collection of Rainfall Data

Daily precipitation data from 1 January 1981 to 31 December 2015 for a total of 38 stations from the National Meteorological Information Center of the China Meteorological Administration were used in the present study [54]. The altitude of these stations varies from 2327 to 4800 m, with the lowest elevation at Zayu station and the highest at Amdo station. The northernmost and westernmost station was Shiquanhe, the southernmost was Pagri, and the easternmost was Markam station (Figure 1).

2.3. Methodology

2.3.1. Calculation of Rainfall Erosivity

Rainfall data were checked for quality. The cold and warm season rainfall estimation model was selected to calculate rainfall erosivity [34,55]. The model takes into full account the characteristics of rainfall in China; that is, the warm season (May–September) with multi-convective, short-duration heavy rain, and the cold season (October–December, January–April) with many frontal, long-duration and heavy rains. The model has been widely used, especially in the first national water census in China [11].

The rainfall erosivity model for estimating the daily rainfall during the warm and cold seasons has the following form:

$$\overline{R} = \sum_{k=1}^{24} \overline{R}_{half\,month\,k}, \tag{1}$$

$$\overline{R}_{half\,month\,k} = \frac{1}{n}\sum_{i=1}^{n}\sum_{j=0}^{m}\left(\alpha \times p_{i,j,k}^{1.7265}\right), \tag{2}$$

$$\overline{WR}_{half\,month\,k} = \frac{\overline{R}_{half\,month\,k}}{\overline{R}}, \tag{3}$$

where $\overline{R}$ is the average annual rainfall erosivity in MJ·mm·ha^{-1}·h^{-1}; k = 1, 2, ... , 24, which divide the year into 24 half months; $\overline{R}_{half\,month\,k}$ is rain erosivity of the k_{th} half-month in MJ·mm·ha^{-1}·h^{-1}; I = 1, 2, ... , n, and n indicates the rainfall data year; j = 0,1, ... , m, m is the number of the erosion-causing rainy days in the k_{th} half month of the i_{th} year; $p_{i,j,k}$ is the j amount of erosive daily rainfall (rainfall ≥12 mm) in the k_{th} month of the i_{th} year in mm, with j = 0 indicating no erosive rainfall resulting in $p_{i,j,k} = 0$; in the dry season α is 0.3101, and in the rainy season α is 0.3937; and $\overline{WR}_{half\,month\,k}$ is the ratio of the average rainfall erosivity of the k_{th} month to the average annual rainfall erosivity.

2.3.2. Trend Coefficient and Climate Tendency Rate

To determine temporal variability in rainfall erosivity, we applied the trend coefficient (TC) and climate tendency rate (CTR) indicators, which are widely used to detect and assess, respectively, the direction and extent of long-term change in climate factors [56,57]. The formula for the trend coefficient is as follows:

$$r_{xt} = \frac{\sum\limits_{i=1}^{n}(x_i - \overline{x})(i - \overline{t})}{\sqrt{\sum\limits_{i}^{n}(x_i - \overline{x})^2 \sum\limits_{i=1}^{n}(i - \overline{t})^2}} \tag{4}$$

$$\overline{t} = (n+1)/2 \tag{5}$$

where r_{xt} is the trend coefficient and n is the number of years. x_i is rainfall erosivity of the i-th year. $\overline{x}$ is the average annual rainfall erosivity for the n years. When the trend coefficient r_{xt} is positive, it indicates a linearly increasing trend of rainfall erosivity, and a negative value indicates a linearly decreasing trend.

A linear equation is usually used to indicate trends in meteorological elements [58,59]. The climate tendency rate was calculated as follows:

$$\widehat{p}_t = a_0 + a_1 t \qquad t = 1, 2, \cdots, n, \tag{6}$$

$$\frac{d\widehat{x}_t}{dt} = a_1, \tag{7}$$

where $a_1 \times 10$ is the climate tendency rate over a 10-year period (MJ·mm·ha^{-1}·h^{-1}·10a^{-1}), indicating the variation of rainfall erosivity per 10 years.

According to the linear regression theory,

$$a_1 = r_{xt}\frac{\sigma_x}{\sigma_t}, \tag{8}$$

where σ_x is the mean square error of the rainfall erosivity and σ_t is the mean square error of sequence 1, 2, ... , n.

2.3.3. Mann–Kendall Trend Test

The Mann–Kendall test [60,61] was used to detect the significance level and the abrupt-change point of the long-term variability in R factors; the specific calculation method was as follows:

Construct a matrix for the time series x which has a sample of n:

$$s_k = \sum_{i=1}^{k} r_i \quad (k = 2, 3, \ldots, n), \tag{9}$$

$$r_i = \begin{cases} +1 & when\, x_i > x_j \\ 0 & otherwise \end{cases} \quad (j = 1, 2, \ldots, i). \tag{10}$$

Define the statistic UF_k given the assumption that the time series is random and independent:

$$UF_k = \frac{[s_k - E(s_k)]}{\sqrt{Var(s_k)}} \quad (k = 1, 2, \ldots, n), \tag{11}$$

where $UF_1 = 0$, and $E(s_k)$ and $Var(s_k)$ are the average and variance of the cumulative number s_k, respectively. When $x1, x2, \ldots, xn$ are mutually independent and of the same continuous distribution, they can be calculated from the equations below:

$$E_{(s_k)} = \frac{n(n+1)}{4}, \tag{12}$$

$$Var(s_k) = \frac{n(n-1)(2n+5)}{72}. \tag{13}$$

UF_i is the standard normal distribution, which is calculated according to the time sequence $x1, x2, \ldots, xn$. Given the significance level α, when$|UF_i| > U\alpha$, there is a significant trend in the sequence.

Repeat the calculation to the inverted sequence $xn, xn-1, \ldots, x1$, and make $UB_k = -UF_k$ ($k = n$, $n - 1, \ldots 1$), $UB = 0$. Intersection of UF and UB is the abrupt change point.

2.3.4. Sen's Slope Estimator

Sen (1968) [62] developed the non-parametric procedure for estimating the slope of the trend in the sample of N pairs of data:

$$Q_i = \frac{x_j - x_k}{j - k} \quad for\, i = 1, \ldots, N, \tag{14}$$

where x_j and x_k are the data values at times j and k ($j > k$), respectively.

If there is only one datum in each time period, then $N = n(n-1)/2$, where n is the number of time periods. If there are multiple observations in one or more time periods, then $N < n(n-1)/2$, where n is the total number of observations. The N values of Q_i are ranked from smallest to largest and the median of slope or Sen's slope estimator is computed as shown:

$$Q_{med} = \begin{cases} Q_{[(N+1)/2]} & if\, N\, is\, odd \\ \frac{Q_{[N/2]} + Q_{[(N+2)/2]}}{2} & if\, N\, is\, even \end{cases}. \tag{15}$$

The sign of Q_{med} reflects the direction of data trend, while its value indicates the steepness of the trend.

The confidence interval about the time slope [62,63] can be computed as follows:

$$C_\alpha = Z_{1-\alpha/2}\sqrt{Var(S)}, \tag{16}$$

where *Var (S)* is defined in Equation (13) and $Z_{1-\alpha/2}$ is obtained from the standard normal distribution table. In this study, the confidence interval was computed at two significance levels (α = 0.01 and α = 0.05).

Then, $M_1 = (N - C_\alpha)/2$ and $M_2 = (N + C_\alpha)/2$ are calculated. The lower and upper limits of the confidence interval, Q_{min} and Q_{max}, are the M_1 largest and the $(M_{2+1})_{th}$ largest of the N ordered slope estimates [63].

The slope Q_{med} is statistically different than zero if the two limits (Q_{min} and Q_{max}) have the same sign.

Both the Mann–Kendall statistical test and Sen's slope have been frequently used to detect the significance of trends in hydro-meteorological time series [64–69].

3. Results

3.1. Temporal Variability in Rainfall Erosivity

Sen's slope estimator analysis indicated that rainfall erosivity in the Tibetan Plateau exhibited an increasing trend from 1980 to 2015, but the trend was not significant. Sen's slope of annual rainfall erosivity was 3.05 ($p = 0.32$). However, variability at each station was high, and 13 stations exhibited a decreasing trend (34% of the stations), and 25 stations an increasing trend (66% of the stations) (Figure 2a, Table 1). The climate tendency rate had a significantly increasing trend at the Gerze station in the northern part of the Tibetan Plateau, Shenza station in the middle, and Markam station in the eastern part of the Minjiang River basin (at the 0.05 significance level), and the rate increased by 20 MJ·mm·ha^{-1}·h^{-1}·10a^{-1}, 32 MJ·mm·ha^{-1}·h^{-1}·10a^{-1}, and 50 MJ·mm·ha^{-1}·h^{-1}·10a^{-1}, respectively, at the three stations. Increasing rainfall erosivity was mainly found in the southern Tibetan valley, Yarlung Zangbo River basin, and Hengduan Mountains in the Tibetan Plateau, especially in the Lancang River Basin. An annual rainfall erosivity anomaly showed that it fluctuated. The rainfall erosivity was small before 1987, and the value increased after 2010. In the study period, two-thirds of the annual rainfall erosivity was higher than the average; the minimum appeared in 1992, and the maximum appeared in 1998 (Figure 2b). The results of the M–K test showed a non-significant increasing trend in rainfall erosivity in the Tibetan Plateau since 1984 (Figure 2c).

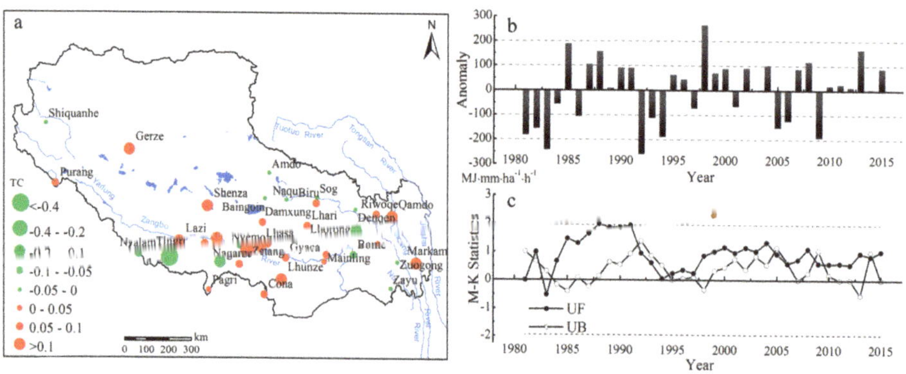

Figure 2. The trend coefficient (**a**), anomalies (**b**), and M–K test (**c**) for annual rainfall erosivity from 1981 to 2015 in the Tibetan Plateau. Note: Dots are scaled according to trend magnitude. The red color corresponds to an increasing trend, and the green color corresponds to a decreasing trend. $UF > 0$ indicates an increasing trend, $UF < 0$ indicates a decreasing trend. The mutation year exists at the intersection of *UF* and *UB*.

Table 1. Statistical tests for annual rainfall erosivity for the period 1981–2015.

Station	Z_s	Q_{med}	Station	Z_s	Q_{med}	Station	Z_s	Q_{med}	Station	Z_s	Q_{med}
Shiquanhe	−0.493	0.000	Shigatse	0.573	5.495	Cona	1.184	4.227	Bome	−0.398	−3.866
Gerze	2.075 *	5.030	Nyemo	−0.308	−1.668	Lhunze	1.71	5.334	Baxoi	1.008	2.447
Baingoin	−0.059	−0.151	Konggar	1.462	8.526	Pagri	0.741	3.367	Gyaca	1.184	7.970
Amdo	−0.294	−1.144	Lhasa	1.641	11.284	Sog	−0.312	−1.725	Nyingchi	−0.068	−0.836
Naqu	0.000	−0.060	Maizhokunggar	0.859	7.607	Biru	0.593	2.892	Mainling	0.652	5.247
Purang	−0.243	−0.502	Zetang	1.564	8.255	Denqen	0.015	0.302	Zuogong	−0.089	−0.184
Shenza	2.102 *	7.084	Nyalam	−0.267	−7.061	Riwoqe	0.741	4.040	Markam	1.957	12.316
Damxung	0.759	3.489	Tingri	1.038	5.141	Qamdo	1.392	6.841	Zayu	−0.625	−4.515
Lazi	1.789	6.831	Gyangze	−1.379	−4.114	Lhari	1.476	5.079			
Nanmulin	1.071	9.389	Nagarze	1.022	5.305	Lhorong	−0.728	−2.851			

Note: Z_s: Mann–Kendall test, Q_{med}: Sen's slope estimator, *: Statistically significant trend at the 0.05 significance level.

The distribution of rainfall erosivity over the 24 and a half months is an input factor for many soil erosion model calculations, including the Chinese Soil Loss Equation (CSLE) [70] and for calculating vegetation cover and biological measure factors. Data from 1981 to 2010 indicated that rainfall erosivity in the Tibetan Plateau was mainly concentrated in June–September, accounting for 81% of the year (Figure 3). This was closely related to the monsoon climate of the Tibetan Plateau, as monsoons account for 58.5% of the annual precipitation [52]. The proportion of rainfall in May–October accounted for 90% of the rain in the eastern part of Tibet [49]. Therefore, the rainy season was also a frequent period of soil water erosion.

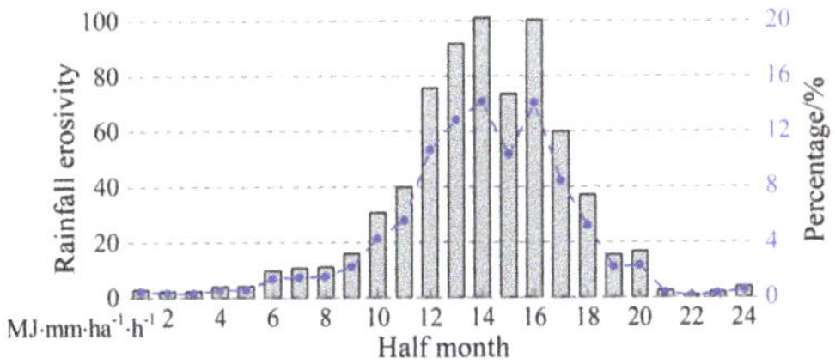

Figure 3. Statistics of half-monthly rainfall erosivity and its percentages.

Rainfall erosivity at most meteorological stations in spring showed an increasing trend; only 5 of the 38 meteorological stations had a decreasing trend (Figure 4a, Table 2). The climate tendency rate indicated a significant increasing trend (at the 0.05 significance level) at Cona and Nyingchi stations in the southern part of the Tibetan Plateau, and a highly significant increase (at 0.01 significance level) at the Sog station in the upper reaches of the Nu River Basin; the climate tendency rate increased by 10, 20, and 11 MJ·mm·ha^{-1}·h^{-1}·10a^{-1}, respectively, for the three stations. Sen's slope analysis indicated that rainfall erosivity in the Tibetan Plateau increased during the study period. Sen's slope value for spring rainfall erosivity was 1.85 ($p = 0.005 < 0.01$). Rainfall erosivity with a decreasing trend was found in the western part of the Tibetan Plateau where the Zuogong station had an insignificant decreasing trend; its erosivity decreased by 40 MJ·mm·ha^{-1}·h^{-1}·10a^{-1} (Table 2). A spring rainfall erosivity anomaly showed that rainfall erosivity was smaller before 2000, and larger after 2000. During the study period, spring rainfall erosivity was below average 67% of the time; the minimum appeared in 1987 and the maximum in 2010 (Figure 4b). The M–K test results showed an increasing trend in spring erosivity in the Tibetan Plateau since 1987, a significant increasing trend after 2010, and the abrupt year occurred in 2005 (Figure 4c).

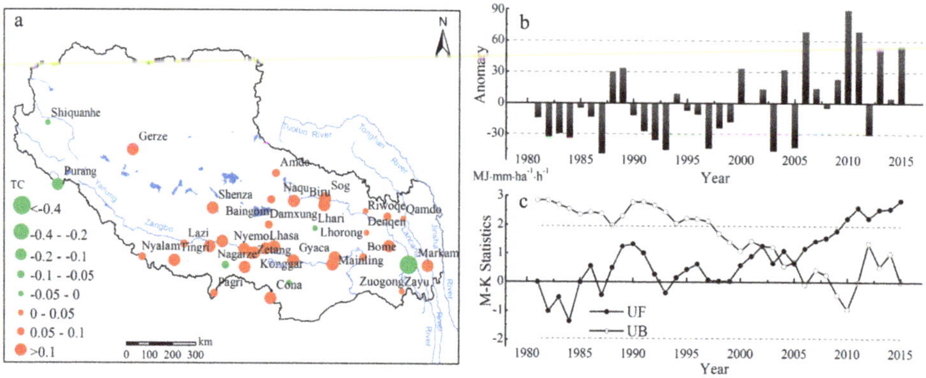

Figure 4. The trend coefficient (**a**), anomalies (**b**), and M–K test (**c**) for spring rainfall erosivity from 1981 to 2015 in Tibetan Plateau. Note: Dots are scaled according to trend magnitude. The red color corresponds to an increasing trend, and the green color corresponds to a decreasing trend. $UF > 0$ indicates an increasing trend, $UF < 0$ indicates a decreasing trend. The mutation year exists at the intersection of UF and UB.

Table 2. Statistical tests for spring rainfall erosivity for the period 1981–2015.

Station	Z_s	Q_{med}	Station	Z_s	Q_{med}	Station	Z_s	Q_{med}	Station	Z_s	Q_{med}
Shiquanhe	-	-	Shigatse	0.615	0	Cona	2.25 *	1.927	Bome	1.449	7.632
Gerze	1.364	0	Nyemo	1.151	0	Lhunze	−0.024	0	Baxoi	1.467	0.485
Baingoin	0.825	0	Konggar	1.044	0	Pagri	1.819	1.512	Gyaca	1.715	0.873
Amdo	0.830	0	Lhasa	1.227	0	Sog	2.753 **	1.966	Nyingchi	2.261 *	3.985
Naqu	1.943	0	Maizhokunggar	−0.066	0	Biru	0.959	0.452	Mainling	1.853	2.571
Purang	−1.198	−0.852	Zetang	0.299	0	Denqen	0.187	0	Zuogong	1.386	0
Shenza	1.913	0	Nyalam	−0.460	−0.950	Riwoqe	0.730	0.332	Markam	1.782	0.815
Damxung	0.882	0	Tingri	1.623	0	Qamdo	0.539	0.133	Zayu	−0.199	−2.253
Lazi	0.917	0	Gyangze	−0.228	0	Lhari	0.130	0.312			
Nanmulin	1.970	0	Nagarze	0.370	0	Lhorong	0.779	0.944			

Note: Z_s: Mann–Kendall test, Q_{med}: Sen's slope estimator, *: Statistically significant trend at the 0.05 significance level, **: Statistically significant trend at the 0.01 significance level.

Increasing rainfall erosivity in summer was observed at more stations than did decreasing, with 21 of 38 meteorological stations (55%) showing increasing trends in rainfall erosivity (Figure 5a, Table 3). Based on the climate tendency rate, Shenza station in the middle of the Tibetan Plateau, and the Qamdo station of the Lancang River Basin in the eastern Hengduan Mountains showed a significant trend (at 0.05 significance level) with an increase of 31 and 41 MJ·mm·ha^{-1}·h^{-1}·10a^{-1}, respectively; the trend at Gerze station in the west and Markam station in the Lancang River Basin in the eastern Hengduan Mountains was highly significant (at the 0.01 significance level) at 24 and 69 MJ·mm·ha^{-1}·h^{-1}·10a^{-1}, respectively. The results of Sen's slope estimator indicated an increasing but not significant trend during the study period. The Sen's slope value of summer rainfall erosivity was 1.41 ($p = 0.41$). Rainfall erosivity showing a decreasing trend was observed mainly in the western part of the Tibetan Plateau, the southern part, and the western part of Hengduan Mountains, with the largest decrease at Gyangze and Lhorong stations where the trend coefficient decreased by 34 and 24 MJ·mm·ha^{-1}·h^{-1}·10a^{-1}, respectively (Table 3). A summer rainfall erosivity anomaly showed that summer rainfall erosivity was above average in most years (57%), with the minimum in 1983 and the maximum in 1998 (Figure 5b). The M–K test results indicated an insignificant increasing trend of summer rainfall erosivity in the Tibetan Plateau since 1983 (Figure 5c).

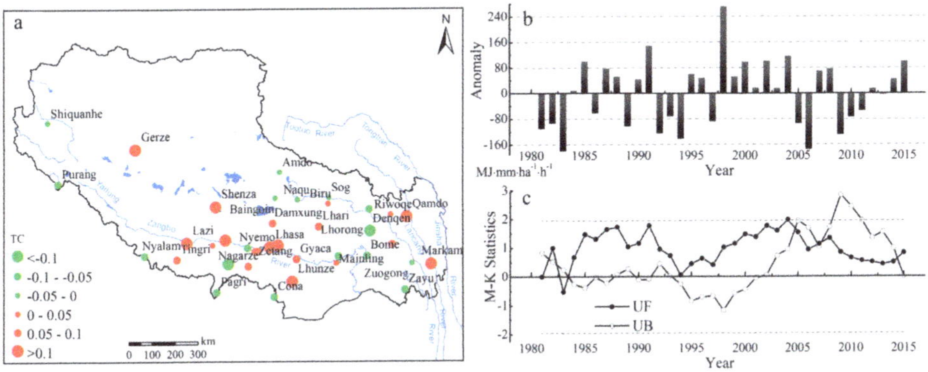

Figure 5. The trend coefficient (**a**), anomalies (**b**), and M–K test (**c**) for summer rainfall erosivity from 1981 to 2015 in the Tibetan Plateau. Note: Dots are scaled according to trend magnitude. The red color corresponds to an increasing trend, and the green color corresponds to a decreasing trend. $UF > 0$ indicates an increasing trend, $UF < 0$ indicates a decreasing trend. The mutation year exists at the intersection of UF and UB.

Table 3. Statistical tests for summer rainfall erosivity for the period 1981–2015.

Station	Z_s	Q_{med}	Station	Z_s	Q_{med}	Station	Z_s	Q_{med}	Station	Z_s	Q_{med}
Shiquanhe	−0.199	−2.253	Shigatse	0.604	5.805	Cona	−0.600	−0.873	Bome	−1.647	-10.265
Gerze	2.861 **	4.857	Nyemo	−0.730	−2.692	Lhunze	1.142	2.823	Baxoi	1.275	2.085
Baingoin	0.208	0.608	Konggar	0.442	3.397	Pagri	−0.504	−1.199	Gyaca	0.795	4.846
Amdo	−0.418	−1.626	Lhasa	2.141	11.923	Sog	−0.483	−1.821	Nyingchi	−0.204	−1.231
Naqu	−0.036	−0.443	Maizhokunggar	0.697	6.002	Biru	−0.089	−0.151	Mainling	0.296	2.119
Purang	−1.262	0	Zetang	1.190	4.986	Denqen	−0.666	−3.152	Zuogong	−0.059	−0.157
Shenza	2.017 *	5.683	Nyalam	−1.008	−3.273	Riwoqe	0.148	0.469	Markam	2.580 **	14.369
Damxung	0.821	2.952	Tingri	0.534	1.774	Qamdo	2.070 *	8.620	Zayu	−1.221	−4.697
Lazi	1.591	5.814	Gyangze	−1.286	−4.041	Lhari	0.697	2.741			
Nanmulin	1.463	11.478	Nagarze	0.924	3.610	Lhorong	−0.976	−2.554			

Note: Z_s: Mann–Kendall test, Q_{med}: Sen's slope estimator, *: Statistically significant trend at the 0.05 significance level, **: Statistically significant trend at the 0.01significance level.

The number of stations with increasing rainfall erosivity in autumn was slightly lower than that with a decreasing trend; 17 of the 38 meteorological stations (45%) showed an increasing trend for rainfall erosivity (Figure 6a, Table 4). Bases on the climate tendency rate, the Konggar station in the Yarlung Zangbo River valley in the southern part of the Tibetan Plateau (near Zetang station) showed a significant increasing trend (at the 0.05 significance level), and the Lhunze station showed a highly significant increasing trend (at the 0.01 significance level), with an increase of 16 and 13 MJ·mm·ha^{-1}·h^{-1}·10a^{-1}, respectively. The results of autumn rainfall erosivity indicated an insignificant decreasing trend during the study period. The Sen's slope value of autumn rainfall erosivity was −0.02 ($p = 0.98$). In addition to the Yarlung Zangbo River Basin, autumn rainfall erosivity in other areas was mainly decreasing (Table 3). Autumn rainfall erosivity at the Bome and Zuogong stations in the Hengduan Mountains in the eastern part of the Tibetan Plateau decreased by 46 and 9 MJ·mm·ha^{-1}·h^{-1}·10a^{-1}, respectively. An autumn rainfall erosivity anomaly shows that autumn rainfall erosivity was above average in most years (50%), with the minimum in 1992 and the maximum in 1985 (Figure 6b). The M–K test results indicated an insignificant increasing trend in the Tibetan Plateau from 1984 to 2003, and a slowly increasing trend after 2008 (Figure 5c).

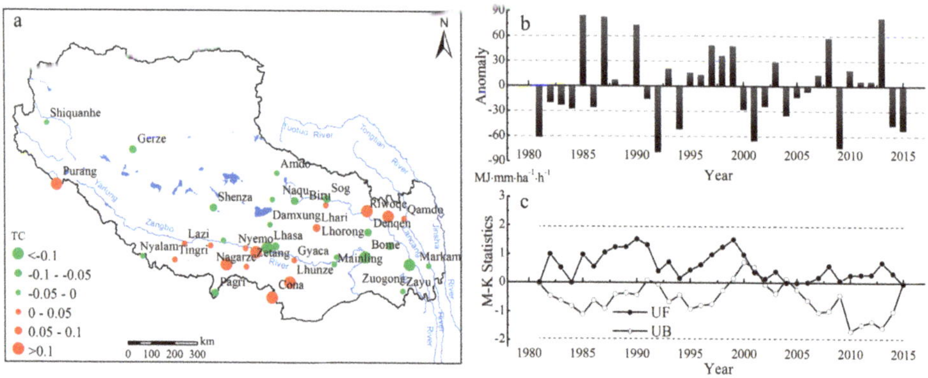

Figure 6. The trend coefficient (**a**), anomalies (**b**), and M–K test (**c**) for autumn rainfall erosivity from 1981 to 2015 in Tibetan Plateau. Note: Dots are scaled according to trend magnitude. The red color corresponds to an increasing trend, and the green color corresponds to a decreasing trend. $UF > 0$ indicates an increasing trend, $UF < 0$ indicates a decreasing trend. The mutation year exists at the intersection of UF and UB.

Table 4. Statistical tests for autumn rainfall erosivity for the period 1981–2015.

Station	Z_s	Q_{med}	Station	Z_s	Q_{med}	Station	Z_s	Q_{med}	Station	Z_s	Q_{med}
Shiquanhe	−0.226	0	Shigatse	0.966	0.934	Cona	1.088	1.124	Bome	−0.824	−3.893
Gerze	−0.233	0	Nyemo	0.309	0.191	Lhunze	3.061 **	2.567	Baxoi	−1.168	−0.775
Baingoin	0.301	0	Konggar	1.990 *	3.093	Pagri	0.58	0.302	Gyaca	−0.016	0
Amdo	−0.171	−0.074	Lhasa	−1.089	−1.582	Sog	−0.824	−0.958	Nyingchi	−0.272	−1.306
Naqu	−0.965	−1.680	Maizhokunggar	−0.730	−1.911	Biru	0.252	0.494	Mainling	−0.682	−1.031
Purang	0.904	0	Zetang	0.034	0	Denqen	0.821	1.740	Zuogong	−1.380	−1.870
Shenza	−0.868	0	Nyalam	−0.326	−2.011	Riwoqe	1.453	2.589	Markam	−0.830	−1.821
Damxung	−0.543	−0.981	Tingri	−0.171	0	Qamdo	0.500	0.753	Zayu	−0.939	−3.321
Lazi	0.432	0	Gyangze	1.238	0	Lhari	1.022	1.779			
Nanmulin	−0.018	0	Nagarze	0.049	0	Lhorong	−0.744	−1.360			

Note: Z_s: Mann–Kendall test, Q_{med}: Sen's slope estimator, *: Statistically significant trend at the 0.05 significance level, **: Statistically significant trend at the 0.01significance level.

Winter rainfall erosivity in the Tibetan Plateau was very low, and about 66% of the stations had a value of 0 (Figure 7a, Table 5). The results indicated an insignificant decreasing trend in winter rainfall erosivity during the study period. Meanwhile, the Sen's slope value of winter rainfall erosivity was −0.03 ($p = 0.89$). Based on the climate tendency rate, the largest decreasing trend was observed at the Lhunze station and Qamdo in the Hengduan Mountains in the eastern part of the Tibetan Plateau, as well as at the Cona station in the southern region, where it decreased almost 100%. The anomaly showed that winter rainfall erosivity was below average in most years (66%), with the minimum in 1992 and the maximum in 198[illegible] (Figure 7b). The M–K test results demonstrated that the winter rainfall erosivity in the Tibetan Plateau had an insignificant decreasing trend since 1991 (Figure 7c).

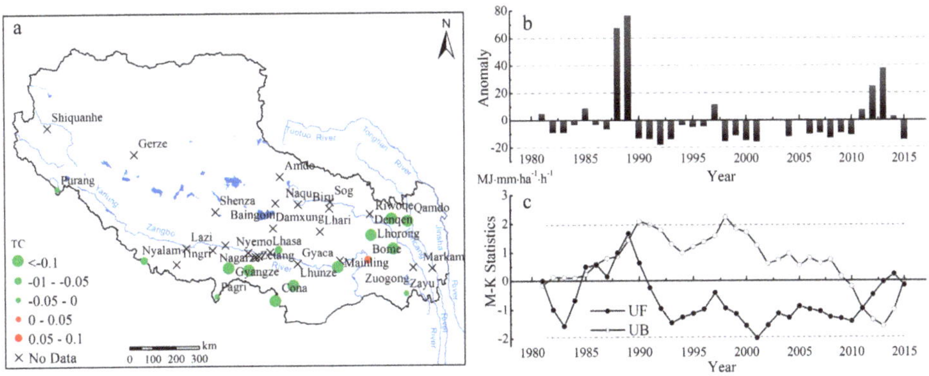

Figure 7. The trend coefficient (**a**), anomalies (**b**), and M-K test (**c**) for winter rainfall erosivity from 1981 to 2015 in Tibetan Plateau. Note: Dots are scaled according to trend magnitude. The red color corresponds to an increasing trend, and the green color corresponds to a decreasing trend. $UF > 0$ indicates an increasing trend, $UF < 0$ indicates a decreasing trend. The mutation year exists at the intersection of UF and UB.

Table 5. Statistical tests for winter rainfall erosivity for the period 1981–2015.

Station	Z_s	Q_{med}	Station	Z_s	Q_{med}	Station	Z_s	Q_{med}	Station	Z_s	Q_{med}
Shiquanhe	-	-	Shigatse	-	-	Cona	−1.564	0	Bome	1.255	0
Gerze	-	-	Nyemo	-	-	Lhunze	−1.618	0	Baxoi	−1.631	0
Baingoin	-	-	Konggar	-	-	Pagri	0.555	0	Gyaca	-	-
Amdo	-	-	Lhasa	-	-	Sog	-	-	Nyingchi	-	-
Naqu	-	-	Maizhokunggar	−0.545	0	Biru	-	-	Mainling	−1.359	0
Purang	0	0	Zetang	-	-	Denqen	-	-	Zuogong	-	-
Shenza	-	-	Nyalam	−0.148	−1.017	Riwoqe	−1.631	0	Markam	-	-
Damxung	-	-	Tingri	-	-	Qamdo	−1.618	0	Zayu	−0.288	0
Lazi	-	-	Gyangze	−1.628	0	Lhari	-	-			
Nanmulin	-	-	Nagarze	−1.625	0	Lhorong	−1.628	0			

Note: Z_S: Mann–Kendall test, Q_{med}: Sen's slope estimator, -: No data.

3.2. Spatial Distribution of Rainfall Erosivity in the Tibetan Plateau

Spatial distribution of rainfall erosivity in the Tibetan Plateau showed a decreasing trend from the southeast to the northwest (Figure 8). The average annual rainfall erosivity at 38 meteorological stations varied from 61 to 1776 MJ·mm·ha^{-1}·h^{-1} during the study period, with an average of 714 MJ·mm·ha^{-1}·h^{-1}. The largest annual rainfall erosivity was observed at the Bome station in the eastern part of the Tibetan Plateau, and the smallest at the Shiquanhe station in the western region (Figure 8). Rainfall erosivity was <500 MJ·mm·ha^{-1}·h^{-1} at 24% of the stations, 500–1000 MJ·mm·ha^{-1}·h^{-1} at 55%, and >1000 MJ·mm·ha^{-1}·h^{-1} at 21% of the stations. Relatively high rainfall erosivity was mainly distributed in the Hengduan Mountains in the eastern part of the Tibetan Plateau and in the low-elevation areas between the Yarlung Zangbo and Nu rivers.

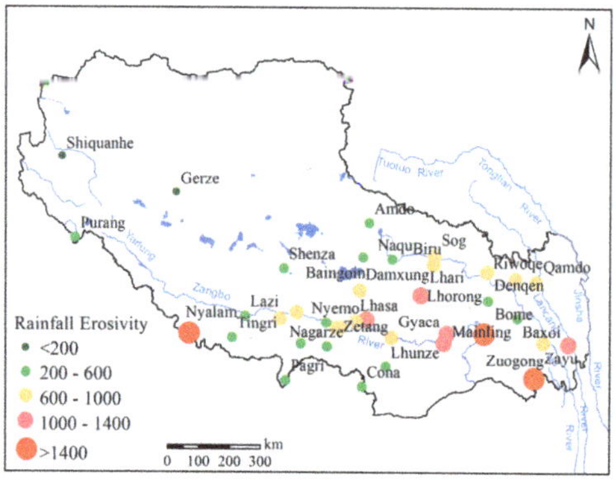

Figure 8. Spatial distribution of annual average rainfall erosivity (MJ·mm·ha^{-1}·h^{-1}) from 1981 to 2015.

The seasonal distribution of rainfall erosivity varied greatly across the Tibetan Plateau; the average summer rainfall erosivity was 482 MJ·mm·ha^{-1}·h^{-1}, accounting for a maximum of 67.5% of the annual rainfall erosivity, followed by autumn and spring, accounting for 18.5% and 11.5%, respectively. The proportion of winter rainfall erosivity was the smallest at only 2.5%. Seasonal variability in rainfall erosivity varied among meteorological stations, but generally followed rain distribution in summer > autumn > spring > winter (Figure 9). The proportions of spring to annual total rainfall erosivity differed among meteorological stations. The Shiquanhe station had the smallest proportion at 0 and the Zayu station in the Hengduan Mountains had a proportion of 46.5%. Meanwhile, the proportion in the south was higher than in the north, and higher in the east than in the west, resulting in a slight decreasing trend from the southeast to the northwest (Figure 9a). The proportion of summer rainfall erosivity varied from the smallest of 16.4% at Nyalam, to the largest of 94.7% at Shiquanhe station. In addition, 71% of the stations accounted for more than 70% of summer rainfall erosivity, and summer rainfall erosivity in most meteorological stations in the southern Tibet Valley contributed more than 80% to the total (Figure 9b). The proportion of autumn rainfall erosivity at each meteorological station varied between 5.3% and 39.4% with the highest at Purang station, followed by the Nyalan and Cona stations; these stations were located on the southernmost edge of the Tibetan Plateau (Figure 9c). The proportion of winter rainfall erosivity to the total was small, with less than 4% of the stations contributing >90% of the winter erosivity. In that, the Nyalam station contributed the most at 33.9%, followed by Purang, both located on the southwestern edge of the Tibetan Plateau (Figure 9d).

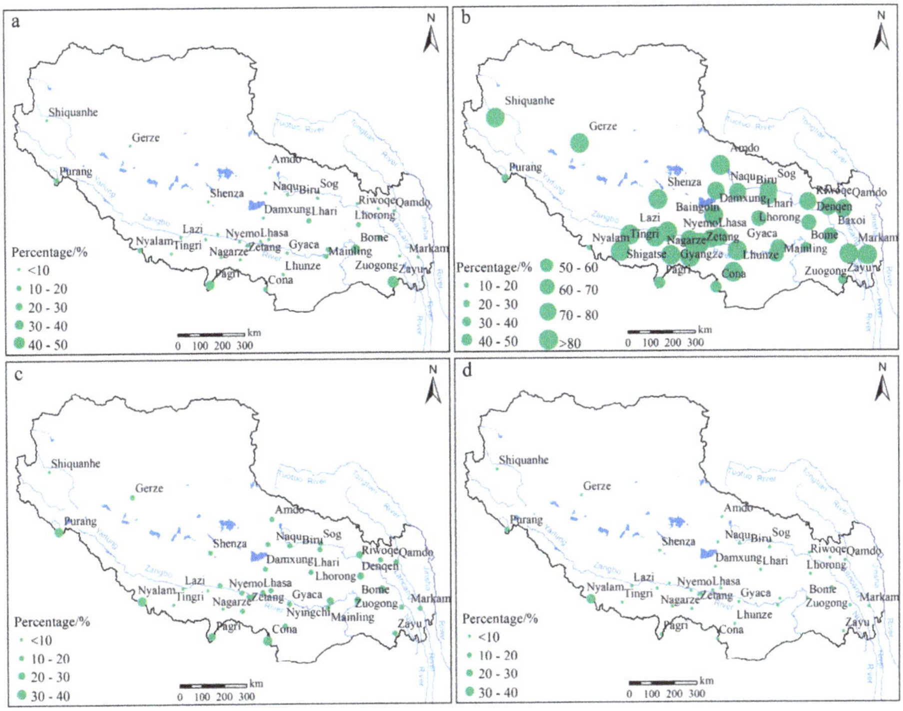

Figure 9. Spatial distribution of spring (**a**), summer (**b**), autumn (**c**), and winter (**d**) rainfall erosivity from 1981 to 2015.

4. Discussion

Rainfall erosivity in the Tibetan Plateau from 1981 to 2015 varied from 61 to 1776 MJ·mm·ha^{-1}·h^{-1}, with an average of 714 MJ·mm·ha^{-1}·h^{-1}. This value is higher than the inland areas of northwest China, but smaller than the areas of east and south China [34]. Yan et al. [48] assessed temporal and spatial distribution of rainfall erosivity in the Tibetan Plateau using TRMM 3B42 data from 2000 to 2008 and showed a decreasing trend during the study period. However, our results showed that from 1980 to 2015, the annual rainfall erosivity indicated an insignificant increase. If we put our research period in 2000–2008, the annual rainfall erosivity also showed a slight decrease. Fan et al. [53] found that rainfall erosivity varied from 32 to 12,189 MJ·mm·ha^{-1}·h^{-1}, with an average of 768 MJ·mm·ha^{-1}·h^{-1} during 2000–2010, which is closed to our findings. Gu et al. [49] also estimated rainfall erosivity in the Hengduan Mountainous Region of Eastern Tibet. Their results showed that rainfall erosivity of Qamdo station based on monthly rainfall was about 300 MJ·mm·ha^{-1}·h^{-1}, which was half of our value. Data precision and models cause the difference. Daily rainfall data is better than monthly rainfall data. Spatial distribution is characterized by a decreasing trend from the southeast to the northwest, which is in accord with previous studies [33,34,50,51].

Topography and monsoon affect rainfall distribution in the study area. The higher the altitude, the less the rainfall (Figure 10a). High altitude and atmospheric circulation patterns have created a cold, dry, and fragile environment, which is very sensitive to soil erosion. Weak rainfall erosivity can cause very severe soil erosion, and terrain has an important influence on the distribution of rainfall erosivity in the Tibetan Plateau. Previous studies have shown that altitude, slope, and aspect affect transport of water vapor and the spatial distribution of precipitation in the study area [71]. The relationship between the average annual rainfall erosivity and altitude of meteorological stations revealed a high correlation coefficient (−0.63), with rainfall erosivity decreasing with increasing altitude (Table 6,

Figure 10b). The main factors affecting rainfall erosivity were rainfall and rainfall intensity. There is a strong correlation between rainfall and rainfall erosivity in the study area (Figure 10c). The correlation coefficient is as high as 0.92.

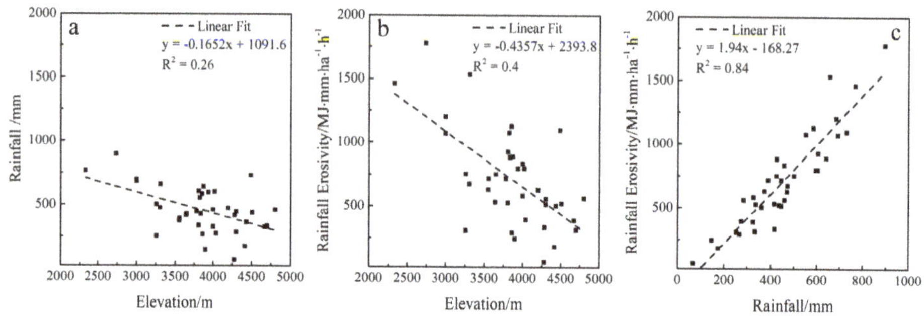

Figure 10. Correlation between elevation and annual rainfall (**a**), elevation and annual rainfall erosivity (**b**), annual rainfall and annual rainfall erosivity (**c**) of meteorological stations.

Table 6. Rainfall (P) and rainfall erosivity (R) for the period 1981–2015.

Station	P	R	Station	P	R	Station	P	R	Station	P	R
Shiquanhe	67	61	Shigatse	425	883	Cona	415	333	Bome	895	1776
Gerze	174	181	Nyemo	336	526	Lhunze	267	292	Baxoi	253	307
Baingoin	333	313	Konggar	389	716	Pagri	445	516	Gyaca	502	753
Amdo	459	561	Lhasa	424	750	Sog	602	797	Nyingchi	684	1201
Naqu	436	521	Maizhokunggar	552	1078	Biru	595	798	Mainling	692	1072
Purang	146	243	Zetang	374	630	Denqen	640	887	Zuogong	444	713
Shenza	325	385	Nyalam	656	1530	Riwoqe	604	924	Markam	585	1127
Damxung	470	626	Tingri	284	557	Qamdo	472	674	Zayu	768	1460
Lazi	326	580	Gyangze	275	389	Lhari	731	1098			
Nanmulin	458	835	Nagarze	361	503	Lhorong	415	532			

The north–south pattern of the mountains and river valleys in Hengduan Mountains in eastern Tibet was conducive to the entry of water vapor from the Indian Ocean, and the rainfall was relatively abundant in that area. Therefore, the annual rainfall erosivity was also relatively high in eastern Tibet, with annual, spring, and summer rainfall erosivity increasing during the study period. In northwestern Tibet, precipitation was extremely small, and the corresponding rainfall erosivity was low; this was due to the obstruction by the mountain system of the Himalayas coupled with high altitude and low temperature. In southern Tibet and in the Himalayas, annual rainfall erosivity was small, mainly due to the local topography of the leeward slope of the mountain. However, in the Yarlung Zangbo river valley, rainfall erosivity was higher due to a lower terrain and sufficient water and heat. Additionally, precipitation in the rainy season has a pronounced vertical differentiation in the plateau, and with the increase in altitude, precipitation significantly decreases, and then increases [71], leading to complexity in the spatial distribution of rainfall erosivity.

Rainfall erosivity in the Tibetan Plateau exhibited a decreasing trend from the southeast to the northwest, due to mainly to the monsoon. Specifically, the trend was affected by the southeast monsoon from the Pacific Ocean, and by the southwest monsoon from the Indian Ocean [72]. The climate of the Tibetan Plateau is diverse, from southeast to northwest, and includes tropical, subtropical, temperate plateau, plateau sub-frigid, and plateau cold zones. The southern Tibetan valley belongs to the subtropical monsoon climate, with relatively abundant precipitation, and relatively high rainfall erosivity. The annual, spring, and summer rainfall erosivity showed an increasing trend in the Tibetan Plateau, with spring erosivity showing an especially significant increase, which was closely related to climate change in this area. The region experienced a warming and more humid trend [73]. A previous study has shown that there was a slight increase in annual rainfall erosivity in most parts of China.

Besides, the central and eastern Qinghai–Tibet Plateau is the region with the most significant increase in rainfall erosivity [33]. Our results also showed that the rainfall erosivity of most meteorological stations on the Tibetan Plateau was increasing. This variation may increase the risk of soil erosion in the region.

The Tibetan Plateau is located in the most active geological and structural tectonic belt with the most varied geological history and the strongest tectonic movement in China; physical weathering of the ground is strong, the freezing and thawing effects are widely distributed, and the material on slopes is unstable, leading to an extremely fragile environment. As shown in this study, the risk of soil erosion in the Hengduan Mountains in the eastern Tibetan Plateau and the Yarlung Zangbo River Valley and its vicinity were high. Rainfall erosivity in this area was not only high, but also showed an increased trend, especially in spring. As the temperature rises, the snow melts and erosion caused by rainfall makes the risk of soil erosion in spring also high. Previous studies also showed that the region with the largest annual erosion modulus in Tibet was the "Three Rivers Basin" in the east (Jinsha River, Lancang River, and Nu River), followed by the Yarlung River valley [74]. The terrain of the Tibetan Plateau is complex; heavy rainfall is likely to cause soil erosion, as well as floods, landslides, and mudslides. Annual rainfall erosivity in the Hengduan Mountains in eastern Tibet was relatively high, and mainly concentrated in summer. The area was high, and with sleep slopes and deep river gorges, and the mountains and valleys coexist. Coupled with flood disasters, it is a region with high risk of soil erosion. Due to the influence of the southwest monsoon in the Indian Ocean, the southern Tibetan valley has excellent hydrothermal conditions and can grow many subtropical crops, but human activities undermine environmental protection, which results in creation of important sources of soil erosion from low-coverage grasslands and sloping farmland in the valley.

5. Conclusions

The Tibetan Plateau is an ecologically fragile area. Climate in the Tibetan Plateau has changed significantly in the early 21st century, especially due to significant warming. The risk of soil erosion brought about by climate change may be significant. In this study, we collected daily precipitation data from 1 January 1981 to 31 December 2015 at 38 meteorological stations in the Tibetan Plateau, and calculated annual and seasonal rainfall erosivity. Temporal and spatial variability patterns of rainfall erosivity were analyzed. We concluded that:

(1) Rainfall erosivity in the Tibetan Plateau from 1981 to 2015 varied from 61 to 1776 MJ·mm·ha^{-1}·h^{-1}, with an average of 714 MJ·mm·ha^{-1}·h^{-1}. The spatial distribution of rainfall erosivity differed significantly across the area, with the value decreasing from the southeast to the northwest.

(2) Summer rainfall erosivity accounted for 67.5% of the total annual rain erosivity, followed by autumn and spring, accounting for 18.5% and 11.5%, respectively. Between 1981 to 2015, and especially since the mid-1980s, annual and summer rainfall erosivity in the Tibetan Plateau exhibited an increasing trend (increased by 50% and 60%); spring erosivity had a highly significant increasing (increased by 94%), while autumn and winter a slightly decreasing trend (less than 10%).

(3) The spatial distribution and temporal variability in rainfall erosivity were affected by the special topography and monsoon of the Tibetan Plateau, as well as by increasing warming and humidification in this area. In contrast, soil erosion risk was high in the Hengduan Mountains in the eastern Tibetan Plateau and in the Yarlung Zangbo River Valley and its vicinity.

Author Contributions: Conceptualization: Z.G., X.D.; methodology: Z.G.; software: T.Y., Z.G.; writing—original draft preparation: Z.G., K.G.; writing—review and editing: Y.L., D.F.; project administration: D.F. All authors have read and agreed to the published version of the manuscript.

Funding: This research was funded by the Second Tibetan Plateau Scientific Expedition and Research Program (STEP) (grant number 2019QZKK0307), and the National Natural Science Foundation of China (grant number 41561063 and 41867029).

Acknowledgments: We are thankful to Beijing Normal University for providing the necessary data and technical support in the rainfall erosivity calculation.

Conflicts of Interest: The authors declare no conflict of interest.

References

1. Pimentel, D. Soil erosion: A food and environmental threat. *Environ. Dev. Sustain.* **2006**, *8*, 119–137. [CrossRef]
2. Smith, P.; House, J.I.; Bustamante, M.; Sobocká, J.; Harper, R.; Pan, G.; West, P.C.; Clark, J.M.; Adhya, T.; Rumpel, C.; et al. Global change pressures on soils from land use and management. *Glob. Chang. Biol.* **2016**, *22*, 1008–1028. [CrossRef]
3. Borrelli, P.; Robinson, D.A.; Fleischer, L.R.; Lugato, E.; Ballabio, C.; Alewell, C.; Meusburger, K.; Modugno, S.; Schütt, B.; Ferro, V.; et al. An assessment of the global impact of 21st century land use change on soil erosion. *Nat. Commun.* **2017**, *8*, 2013. [CrossRef]
4. Wischmeier, W.H.; Smith, D.D. Rainfall energy and its relationship to soil loss. *Trans. AGU* **1958**, *39*, 285–291. [CrossRef]
5. Wischmeier, W.H. A rainfall erosion index for a universal soil loss equation. *Soil Sci. Soc. Am. J.* **1959**, *23*, 246–249. [CrossRef]
6. Hudson, N. *Soil Conservation*; Cornell University Press: Ithaca, NY, USA, 1971.
7. Wischmeier, W.H.; Smith, D.D. *Predicting Rainfall Erosion Losses: A Guide to Conservation Planning*; Agriculture Handbook 537; U.S. Department of Agriculture, Agricultural Research Service: Washington, DC, USA, 1978.
8. Renard, K.G.; Foster, G.R.; Weesies, G.A.; McCool, D.K.; Yoder, D.C. *Predicting Soil Erosion by Water: A Guide to Conservation Planning with the Revised Universal Soil Loss Equation (RUSLE)*; United States Department of Agriculture: Washington, DC, USA, 1997.
9. Kinnell, P.I.A.; Risse, L.M. USLE-M: Empirical modeling rainfall erosion through runoff and sediment concentration. *Soil Sci. Soc. Am. J.* **1998**, *62*, 1667–1672. [CrossRef]
10. Foster, G.R. *User's Reference Guide: Revised Universal Soil Loss Equation (RUSLE2)*; U.S. Department of Agriculture, Agricultural Research Service: Washington, DC, USA, 2004.
11. Liu, B.Y.; Guo, S.Y.; Li, Z.G. China's hydraulic erosion sampling survey. *Soil Water Conserv. China* **2013**, *10*, 26–34. (In Chinese)
12. Panagos, P.; Borrelli, P.; Poesen, J.; Ballabio, C.; Lugato, E.; Meusburger, K.; Montanarella, L.; Alewell, C. The new assessment of soil loss by water erosion in Europe. *Environ. Sci. Policy* **2015**, *54*, 438–447. [CrossRef]
13. Mikhailova, E.A.; Bryant, R.B.; Schwager, S.J.; Smith, S.D. Predicting rainfall erosivity in Honduras. *Soil Sci. Am. J.* **1997**, *61*, 273–279. [CrossRef]
14. Shi, D.M.; Jiang, D.; Lu, X.P.; Jiang, G.Y. Temporal distribution characteristics of rainfall erosivity in Fuling District, Chongqing. *Trans. CSAE* **2008**, *24*, 16–21. (In Chinese)
15. Bonilla, C.A.; Vidal, K.L. Rainfall erosivity in Central Chile. *J. Hydrol.* **2011**, *410*, 126–133. [CrossRef]
16. Lee, J.H.; Heo, J.H. Evaluation of estimation methods for rainfall erosivity based on annual precipitation in Korea. *J. Hydrol.* **2011**, *409*, 30–48. [CrossRef]
17. Alipour, Z.T.; Mahdian, M.H.; Pazira, E.; Hakimkhani, S.; Saeedi, M. The determination of the best rainfall erosivity index for Namak Lake Basin and evaluation of spatial variations. *J. Basic Appl. Sci. Res.* **2012**, *2*, 484–494.
18. Renard, K.G.; Freimund, J.R. Using monthly precipitation data to estimate the R-factor in the revised USLE. *J. Hydrol.* **1994**, *157*, 287–306. [CrossRef]
19. Oduro-Afriyie, K. Rainfall erosivity map for Ghana. *Geoderma* **1996**, *74*, 161–166. [CrossRef]
20. Ferro, V.; Porto, P.; Yu, B. A comparative study of rainfall erosivity estimation for southern Italy and southeastern Australia. *Hydrol. Sci. J.* **1999**, *44*, 3–24. [CrossRef]
21. Richardson, C.W.; Forster, G.R.; Wright, D.A. Estimation of erosion index from daily rainfall amount. *Trans. ASABE* **1983**, *26*, 153–156. [CrossRef]
22. Yu, B.; Rosewell, C.J. An assessment of a daily rainfall erosivity model for New South Wales. *Soil Res.* **1996**, *34*, 139–152. [CrossRef]
23. Zhang, W.B.; Xie, Y.; Liu, B.Y. Rainfall erosivity estimation using daily rainfall amounts. *Sci. Geogr. Sin.* **2002**, *22*, 705–711. (In Chinese)
24. Petkovsek, G.; Mikos, M. Estimating the R factor from daily rainfall data in the sub-Mediterranean climate of southwest Slovenia. *Hydrolog. Sci. J.* **2004**, *49*, 869–877. [CrossRef]
25. Angulo-Martinez, M.; Begueria, S. Estimating rainfall erosivity from daily precipitation records: A comparison among methods using data from the Ebro Basin (NE Spain). *J. Hydrol.* **2009**, *379*, 111–121. [CrossRef]

26. Mannaerts, C.M.; Gabriels, D. Rainfall erosivity in Cape Verde. *Soil Tillage. Res.* **2000**, *55*, 207–212. [CrossRef]
27. Yin, S.Q.; Xie, Y.; Wang, C.G. Calculation of rainfall erosivity by using hourly rainfall data. *Geogr. Res.* **2007**, *26*, 541–547. (In Chinese)
28. Hamlaoui-Moulai, L.; Mesbah, M.; Souag-Gamane, D.; Medjerab, A. Detecting hydro-climatic change using spatiotemporal analysis of rainfall time series in Western Algeria. *Nat. Hazards.* **2013**, *65*, 1293–1311. [CrossRef]
29. Ramos, M.C.; Durán, B. Assessment of rainfall erosivity and its spatial and temporal variabilities: Case study of the Penedès area (NE Spain). *Catena* **2014**, *123*, 135–147. [CrossRef]
30. Silva, A.M. Rainfall erosivity map for Brazil. *Catena* **2004**, *57*, 251–259. [CrossRef]
31. Vallebona, C.; Pellegrino, E.; Frumento, P.; Bonari, E. Temporal trends in extreme rainfall intensity and erosivity in the Mediterranean region: A case study in southern Tuscany, Italy. *Clim. Chang.* **2015**, *128*, 139–151. [CrossRef]
32. Zhang, W.B.; Xie, Y.; Liu, B.Y. Spatial distribution of rainfall erosivity in China. *J. Mt. Sci.* **2003**, *21*, 33–40. (In Chinese)
33. Liu, B.T.; Tao, H.P.; Song, C.F.; Guo, B.; Shi, Z.; Zhang, C.; Kong, B.; He, B. Temporal and spatial variations of rainfall erosivity in China during 1960 to 2009. *Geogr. Res.* **2013**, *32*, 245–256. (In Chinese)
34. Yin, S.Q.; Zhang, W.B.; Xie, Y.; Liu, S.H.; Liu, F. Spatial distribution of rainfall erosivity in China based on high density battle network. *Soil Water Conserv. China* **2013**, *10*, 45–50. (In Chinese)
35. IPCC. Working Group I Contribution to the IPCC Fifth Assessment Report, Climate Change 2013: The Physical Science Basis: Summary for Policymakers. Available online: http://www.climatechange2013.org/report/ (accessed on 28 October 2013).
36. Sun, H.L.; Zheng, D. *The Formation, Evolution and Development of the Qinghai-Tibet Plateau*; Science and Technology Press: Guangzhou, China, 1998. (In Chinese)
37. Sun, K.H.; Zeng, X.D.; Li, F. Climate change characteristics in ecological fragile zones in China during 1980–2014. *Clim. Environ. Res.* **2019**, *24*, 455–468. (In Chinese)
38. Huang, X.L.; Zhang, T.B.; Yi, G.H.; He, D.; Zhou, X.B.; Li, J.J.; Bie, X.J.; Miao, J.Q. Dynamic Changes of NDVI in the Growing Season of the Tibetan Plateau During the Past 17 Years and Its Response to Climate Change. *Int. J. Environ. Res. Public Health* **2019**, *16*, 3452. [CrossRef] [PubMed]
39. Xu, Z.X.; Gong, T.L.; Li, J.Y. Decadal trend of climate in the Tibetan Plateau—Regional temperature and precipitation. *Hydrol. Process.* **2008**, *22*, 3056–3065. [CrossRef]
40. Wang, X.D.; Zhong, X.H.; Fan, J.R. Assessment and spatial distribution of sensitivity of soil erosion in Tibet. *J. Geogr. Sci.* **2004**, *14*, 41–46. [CrossRef]
41. Xu, X.L.; Liu, W.; Kong, Y.P.; Zhang, K.L.; Yu, B.; Chen, J.D. Runoff and water erosion on road side-slopes: Effects of rainfall characteristics and slope length. *Transp. Res. D Transp. Environ.* **2009**, *14*, 497–501. [CrossRef]
42. Gu, S.X. Study on the soil and water loss control model of the typical watershed in the eastern Tibet. *Res. Soil Water Conserv.* **2012**, *19*, 86–89. (In Chinese)
43. Fang, G.L.; Xiang, B.; Du, J.Q.; Wang, B.L.; Zhang, L.K.; Hu, Y.; Xu, J. Risk assessment of non-point source pollution export in Lasahe basin. *TSCAE* **2015**, *31*, 247–254. (In Chinese)
44. Wang, G.X.; Qian, J.; Cheng, G.D.; Lai, Y.M. Soil organic carbon pool of grassland soils on the Qinghai-Tibetan Plateau and its global implication. *Sci. Total Environ.* **2002**, *291*, 207–217.
45. Liu, S.Z.; Fan, J.R.; Zhou, L. Degradation of grass lands and its preventive countermeasures in Tibet Autonomous Region. *Chin. J. Eco-Agric.* **2002**, *10*, 1–3. (In Chinese)
46. Li, Q.; Zhang, C.L.; Shen, Y.P.; Jia, W.R.; Li, J. Developing trend of aeolian desertification in China's Tibet Autonomous Region from 1977 to 2010. *Environ. Earth Sci.* **2016**, *75*, 895. [CrossRef]
47. Shi, Z.; Mao, W.Y. Tibetan water resource and conservancy construction. *Yangtze River* **1999**, *30*, 52–54. (In Chinese)
48. Yan, D.; Fan, J.R.; Guo, F.F.; Guo, X.; Gong, K.F. Spatiotemporal distribution of precipitation erosivity in Tibet Autonomous Region. *Bull. Soil Water Conserv.* **2010**, *30*, 17–21. (In Chinese)
49. Gu, S.X.; Wang, X.D.; Liu, S.Z. The preliminary research on rainfall erosivity based on simple estimation method in the Hengduan Mountainous Region of the eastern Tibet. *Res. Soil Water Conserv.* **2011**, *18*, 28–31. (In Chinese)
50. Wang, Y.S.; Cheng, C.C.; Xie, Y.; Liu, B.Y.; Yin, S.Q.; Liu, Y.N.; Hao, Y.F. Increasing trends in rainfall-runoff erosivity in the Source Region of the Three Rivers, 1961–2012. *Sci. Total Environ.* **2017**, *592*, 639–648. [CrossRef]

51. Panagos, P.; Borreli, P.; Meusburger, K.; Yu, B.; Klik, A.; Lim, K.J.; Yang, J.E.; Niabihi, M.; Larionov, G.A.; Krasnov, S.F.; et al. Global rainfall erosivity assessment based on high-temporal resolution rainfall records. *Sci. Rep.* **2017**, *7*, 4175. [CrossRef] [PubMed]
52. Du, J.; Ma, Y.C. Climate analysis of precipitation trends in Tibetan Plateau. *J. Geogr. Sci.* **2004**, *3*, 56–63. (In Chinese)
53. Fan, J.R.; Chen, Y.; Yan, D.; Guo, F.F. Characteristics of rainfall erosivity based on tropical rainfall measuring mission data in Tibet, China. *J. Mt. Sci.* **2013**, *10*, 1008–1017. [CrossRef]
54. National Meteorological Infomation Center. 2015. Available online: http://data.cma.gov.cn (accessed on 11 January 2019).
55. Xie, Y.; Yin, S.Q.; Liu, B.Y.; Nearing, M.A.; Zhao, Y. Models for estimating daily rainfall erosivity in China. *J. Hydrol.* **2016**, *535*, 547–558. [CrossRef]
56. Liu, C.M.; Zeng, Y. Changes of pan evaporation in the recent 40 years in the Yellow River Basin. *Water Int.* **2004**, *29*, 510–516. [CrossRef]
57. Wu, J.; Fu, C.B.; Zhang, L.Y.; Tang, J.P. Trends of visibility on sunny days in China in the recent 50 years. *Atmos. Environ.* **2012**, *55*, 339–346. [CrossRef]
58. Liu, M.; Peng, X.Y.; Li, J.X.; Pu, M.J. Climatic characteristics of high temperature in Jiangsu Province in recent 50 years. *Resour. Sci.* **2011**, *33*, 1991–1998. (In Chinese)
59. Wang, F.Y.; Liu, P.X. Characteristics analysis on sunshine duration change in Jiuquan Oasis in recent 45 years. *Plateau Meteorol.* **2010**, *29*, 999–1004.
60. Mann, H.B. Nonparametric tests against trend. *Econometrica* **1945**, *13*, 245–259. [CrossRef]
61. Kendall, M.G. *Rank Correlation Methods*; Griffin: London, UK, 1975.
62. Sen, P.K. Estimates of the regression coefficient based on Kendall's tau. *J. Am. Stat. Assoc.* **1968**, *63*, 1379–1389. [CrossRef]
63. Gilbert, R.O. *Statistical Methods for Environmental Pollution Monitoring*; John Wiley and Sons: New York, NY, USA, 1987.
64. Lettenmaier, D.P.; Wood, E.F.; Wallis, J.R. Hydro-climatological trends in the continental United States, 1948–1988. *J. Clim.* **1994**, *7*, 586–607. [CrossRef]
65. Partal, T.; Kahya, E. Trend analysis in Turkish precipitation data. *Hydrol. Process.* **2006**, *20*, 2011–2026. [CrossRef]
66. Modarres, R.; Silva, V.P.R. Rainfall trends in arid and semi-arid regions of Iran. *J. Arid. Environ.* **2007**, *70*, 344–355. [CrossRef]
67. ElNesr, M.N.; Abu-Zreig, M.M.; Alazba, A.A. Temperature trends and distribution in the Arabian Peninsula. *Am. J. Environ. Sci.* **2010**, *6*, 191–203. [CrossRef]
68. Tabari, H.; Marofi, S. Changes of pan evaporation in the west of Iran. *Water Resour. Manag.* **2011**, *25*, 97–111. [CrossRef]
69. Gu, Z.J.; Duan, X.W.; Liu, B.; Hu, J.M.; He, J.N. The spatial distribution and temporal variation of rainfall erosivity in the Yunnan Plateau, Southwest China: 1960–2012. *Catena* **2016**, *145*, 291–300.
70. Liu, B.Y.; Zhang, K.L.; Xie, Y. An empirical soil loss equation. In Proceedings of the 12th International Soil Conservation Organization Conference, Beijing, China, 26–31 May 2002; Tsinghua University Press: Beijing, China, 2002; Volume II, p. 15.
71. Zhang, J.; Li, D.L.; He, J.M.; Wang, X.Y. Influence of terrain on spatial distribution of precipitation in rainy and rainy years of Qinghai-Tibet Plateau. *Adv. Water Sci.* **2007**, *3*, 11–18. (In Chinese)
72. Shu, S.J.; Yu, Z.F.; Wang, Y.; Bai, M. A statistic model for the spatial distribution of precipitation estimation over the Tibetan complex terrain. *China J. Geophys.* **2005**, *48*, 535–542. (In Chinese) [CrossRef]
73. Piao, S.L.; Cui, M.D.; Chen, A.P.; Wang, X.H.; Ciais, P.; Liu, J.; Tang, Y.H. Altitude and temperature dependence of change in the spring vegetation green-up date from 1982 to 2006 in the Qinghai-Xizang Plateau. *Agric. For. Meteorol.* **2011**, *151*, 1599–1608. [CrossRef]
74. Li, D.M. Preliminary Analysis on the Causes, Harm and Difficulties of Soil and Water Loss Distribution in Tibet. *Tibet Sci. Tech.* **2001**, *1*, 21–24. (In Chinese)

water

MDPI

Article

Effect of Rain Peak Morphology on Runoff and Sediment Yield in Miyun Water Source Reserve in China

Jiajia Xu [1,*], Jianjun Zhang [1,*], Minyi Li [2] and Fenzhong Wang [2]

1 School of Soil and Water Conservation, Beijing Forestry University, Beijing 100083, China
2 Soil and Water Conservation Workstation in Miyun District, Beijing 101500, China; minyilimysw@gmail.com (M.L.); wangfenzhongmysw@gmail.com (F.W.)
* Correspondence: xjj_xz@163.com (J.X.); zhangjianjun@bjfu.edu.cn (J.Z.)

Received: 16 September 2019; Accepted: 9 November 2019; Published: 20 November 2019

Abstract: The research on the impact of rainfall patterns on runoff and sediment yield is still insufficient, especially under natural rainfall conditions. We analyzed the influence of rain peak morphology on runoff and sediment yield based on the data of rainfall, runoff, and sediment in the bare runoff plot of Shixia, a small watershed in the Miyun district of Beijing, from 2007 to 2016. We took 0.4 mm min^{-1} as the standard of rain peak classification and the peak width, peak number, peak value, peak position and multi-peak continuity as the indexes of rain peak morphology. The results showed that: (1) Peak number, peak value, and peak width were significantly correlated with runoff and sediment yield, while peak position was irrelevant. The order of correlation between rain peak morphology indexes and runoff yield was peak width (0.71) > peak number (0.69) > peak value (0.33) > peak position (0.05). The order of correlation between rain peak morphological indexes and sediment yield was peak width (0.62) > peak value (0.36) > peak number (0.36) > peak position (−0.09). The multi-peak continuity was not correlated with runoff (0.12) and sediment yield (0.45). (2) When the number of rain peaks was greater than one in a single rainfall, the amount of runoff and sediment production increased significantly. (3) For multi-peak rainfall, 90 min was the boundary point of the rain peak interval, and the sediment yield formed by rainfall with a rain peak continuity >1/90 min^{-1} was significantly larger than the rainfall of ≤1/90 min^{-1}. (4) Covariance analysis showed that the runoff caused by rainfall with a peak at the middle positions was obviously more than rainfall with a peak at the front position. However, the peak position had no significant effect on the sediment yield. (5) The peak rainfall amount of a rainfall (*TPR*) was a comprehensive index reflecting peak number, peak value and peak width, and the correlation between it and the sediment yield and runoff reached 0.60 and 0.71, respectively. Statistical rainfall characteristic indexes included rainfall amount, average rainfall intensity, rainfall duration, I_5 (maximum 5-min rainfall intensity), I_{10}, I_{15}, I_{20}, I_{30}, and I_{60}, among which I_{60} had the strongest correlation with runoff and sediment yield (0.69, 0.60), which were much larger than other rainfall indexes (0.08~0.47, 0.14~0.48) except rainfall amount (0.75, 0.37). By establishing a regression equation, it was found that both *TPR* and I_{60} had good explanatory power for runoff and weak explanatory power for sediment yield.

Keywords: rainfall peak; morphological characteristics; runoff; sediment yield

1. Introduction

Researchers usually use statistical rainfall indicators such as average rainfall intensity and rainfall amount to analyze the relationship between rainfall and soil erosion. Rainfall intensity was found to be the most important factor, the increase of which lead to an increase of runoff and erosion [1–3]. However, many scholars thought that the explanatory power of the above indicators is limited. For example, in the Loess Plateau and Northeast China, the correlation between rainfall amount and soil erosion is very low [4,5], and the maximum rainfall during the period, such as I_{30} (maximum 30-min rainfall intensity), was chosen as the rainfall indicator [6,7].

In previous experimental studies using artificial rainfall, researchers mainly focused on the effects of even rainfall intensity on surface infiltration, runoff, and soil loss [8–10]. However, studies have shown that the internal structure of rainfall processes, i.e., rainfall patterns, had an important influence on infiltration, runoff, and erosion processes [11–13]. Dunkerley [14] scoured dry soil with crusts and without vegetation using instantaneously varied rainfall intensity generated by artificial rainfall experiments. The results showed that the runoff coefficient and flood peak caused by the rainfall process with constant instantaneous rainfall intensity were significantly increased by 570% compared with the rainfall process with even rainfall intensity. In artificial rainfall experiments, if the rainfall pattern cannot be designed to match the rainfall characteristics of the local natural rainfall process, the estimation of soil erosion can be erroneous.

Since the soil erosion process is affected by multi-factor coupling, erosion intensity prediction is a current problem for academics, especially the uncertainty of time scale predictions of individual rainfall. Regardless of which statistical models or physical models are used, the relative errors of related studies are mostly between 20% and 30%, rather than within 10% [15]. Hudson's study [16] in south-central Africa found that a total kinetic energy of individual rainfall stronger than 25 mm/h could be regarded as an index of erosivity, which is called the Kinetic Energy (KE) > 25 method. Huff [17] equally divided the rainfall duration into four periods according to which quarter of the rainfall duration the peak rainfall intensity occurred. All of these studies contributed to the improvement of soil erosion model accuracy.

This paper creatively proposes a morphological index of rainfall peaks to characterize rainfall and to analyze its effects on runoff and sediment yield, which includes peak number, peak value, peak position, peak width, and multi-peak continuity. It is rare to analyze the effect of rainfall on runoff and sediment yield from the point of view of rain peak morphology. Only some studies on peak position are available, as mentioned above. Based on 46 rainfall data points we aimed to: (1) quantitatively analyze the effect of the morphological characteristics of rainfall peak on runoff and sediment yield, (2) compare the advantages and disadvantages of the morphological indexes of rainfall peak with the traditional statistical rainfall indexes, and (3) propose an applied model of runoff and sediment yield based on rain peak morphological indexes.

2. Materials and Methods

2.1. Study Area

The Shixia watershed is located in the northeast of the upper reaches of Miyun reservoir in Beijing, within the primary protection zone of the Miyun reservoir, between 117°01′–117°07′ E and 47°32′–47°38′ N, with a total watershed area of 33 km^2, as shown in Figure 1. The small Shixia watershed is also located in the Yanshan mountain range, in the transition zone from the piedmont alluvial plain to the mountainous region. It is an earth-rock shallow hilly area with high terrain in the north and low terrain in the south, whose altitude is 150–390 m. The gully density is 0.2 km km^{-2}, and the slopes are gentle with a slope grade larger than 20° only accounting for 16.2% of the watershed area. The lithology is dominated by gneiss, among which are scattered granite and limestone. The soil type is brown soil developed on alluvial and diluvial parent materials with light soil texture, deep soil layer, compact soil mass, and shallow tillage layer. The basin has a warm temperate monsoon climate. It is dry and cold in winter with an average temperature in January of −6.6 °C and is hot in summer with an average temperature in July of 25.3 °C. The annual evaporation of the whole basin is 1840 mm, and the annual average rainfall is 661.8 mm. Eighty percent of the precipitation occurs from June to September. The vegetation in the northern part of the basin is dominated by weeds and shrubs. The vegetation in the hilly area is mostly artificial *Robinia pseudoacacia*, *Pinus tabulaeformis*, and economic forest. Human activities have been frequent for a long time, and natural vegetation types, having been destroyed, are hard to see.

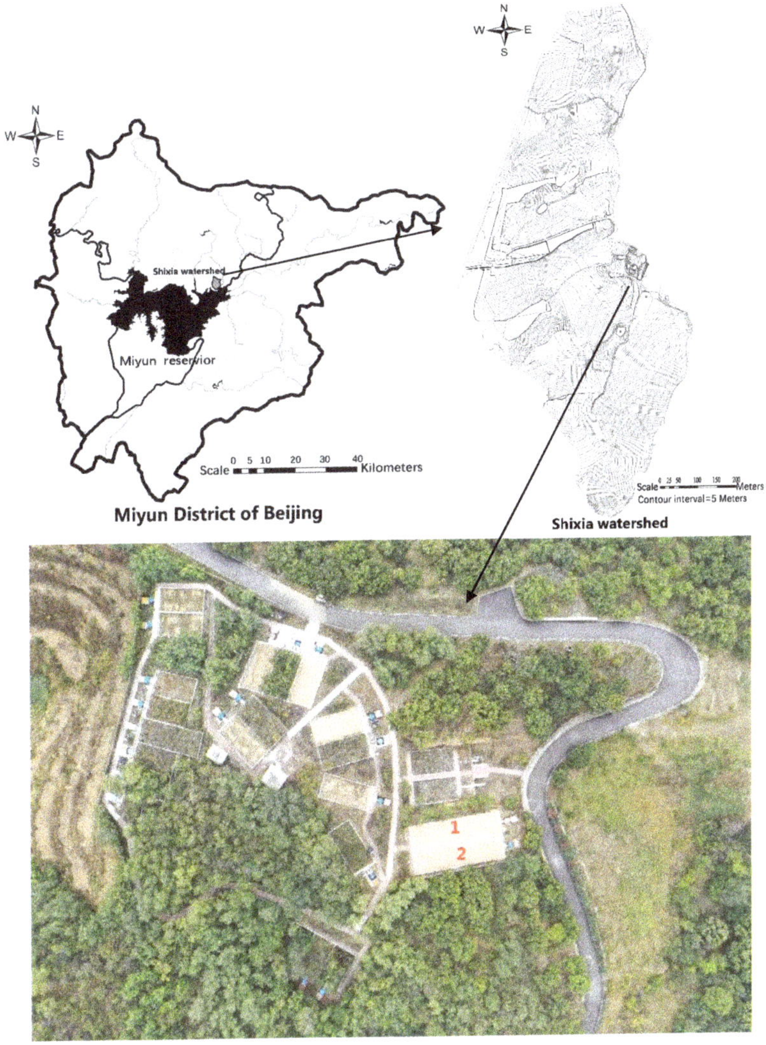

Figure 1. Location of Shixia watershed and runoff plot.

2.2. Sampling Site and Experimental Design

The data in this study were collected from two bare land runoff plots, which are located in the central demonstration area of the Shixia watershed. The two plots both are 20 m long and 5 m wide, with a slope of 14.4°, a slope direction of N270° and 20 cm of soil. Figure 1 is a schematic diagram of the location of the Shixia watershed and runoff plot.

2.3. Rainfall Data Acquisition and Rainfall Characteristic Index Selection

Rainfall was measured by a self-recording rain gauge and standard rain barrel. According to the rainfall records, the rainfall amount, rainfall duration, rainfall intensity, I_5 (maximum 5-min rainfall intensity), I_{10}, I_{15}, I_{20}, I_{30}, and I_{60} were calculated (see Table A1 for data).

In the selection of rainfall characteristic indexes, the morphology of the whole rainfall process needed to be considered. Therefore, on the basis of sorting and analyzing 46 rainfall data points, we put forward rain peak-related indexes to reflect the rainfall characteristics. The minimum peak of rainfall intensity that could cause bare land erosion in this research area was 0.4 mm min^{-1}, so rainfall with a rainfall intensity above 0.4 mm min^{-1} was defined as the rain peak. The definition of specific rain peak morphological indexes were as follows:

Peak number: the number of peaks during a single rainfall.

Peak position: according to the time of the occurrence of the peak during a rainfall, it could be divided into the front position, middle position, and rear position. The rainfall peak is in the front position in 0–30%, in the middle position in 30–60% and in the back position in 60–100% of the rainfall duration. The peak position of a multi-peak rainfall, where the number of rainfall peaks was more than one, was determined by the occurrence time of 40% of the rainfall concentration according to the dividing method of Wang [18]. According to this, rainfall could be divided into front-position rainfall, middle-position rainfall, rear-position rainfall, and uniform rainfall.

Peak value: the highest rainfall intensity. The peak during a single rainfall with the highest value was the highest peak.

Peak width: referred to the duration when the rainfall intensity reached above 0.4 mm min^{-1}.

Peak continuity: the reciprocal of the average interval time between adjacent rainfall peaks. The interval between two adjacent peaks refers to the time from the end of the previous peak to the head of the following peak. This paper attempted to analyze the influence of different multi-peak continuities on runoff and sediment yield, besides the correlation analysis of multi-peak continuity and runoff and sediment yield, the difference of runoff and sediment yield caused by multi-peak continuity within and outside 1/60 min^{-1} and 1/90 min^{-1} were analyzed by variance analysis by taking the multi-peak continuity of 1/60 min^{-1} and 1/90 min^{-1} as dividing nodes.

Peak rainfall: The total peak rainfall amount of one rainfall was referred to as *TPR* for short, and the rainfall amount of the highest peak of a single rainfall was referred to as *HPR* for short.

The data were shown in Table A2.

2.4. Runoff and Erosion Measurements

Runoff and sediment were measured in runoff plots. The volume method was used to measure runoff after each rainfall event. When the amount of slurry was small, 1000 mL of slurry sample was taken after full mixing. When there was more mud, stratified sampling was conducted, and 500 mL was taken in each layer. Runoff and sediment were calculated by dividing the total amount of the two plots by the total area of the two plots (see Table 3 for data).

2.5. Coefficient of Variation

The coefficient of variation (C_v) was used to reflect the inter-annual changes in precipitation, runoff, and sediment. It is calculated using the following equation:

$$C_v = S_d/M, \quad (1)$$

where S_d is the standard deviation, M is the average. The greater the C_v value of precipitation or runoff, which means the greater the variation range of precipitation or runoff, the greater the possibility of flood or drought. The larger the C_v value of sediment yield, which means the greater the variation range of sediment yield, the more common the disasters such as soil erosion are.

2.6. Construction and Test of the Regression Equation

On the basis of Pearson correlation analysis, a regression equation was established by selecting the rainfall characteristic indexes that were most closely related to soil erosion, and the application accuracy of the regression equation was determined by cross-examination. The accuracy evaluation indexes include the mean relative error (*MRE*), coefficients of determination (R^2) and Nash–Suttclife efficiency coefficient (E_{ns}).

$$MRE = \frac{1}{n}\sum_{i=1}^{n}\left|\frac{O_i - Y_i}{O_i}\right|, \tag{2}$$

$$R^2 = \frac{\left(\sum_{i=1}^{n}(O_i-\overline{O})(Y_i-\overline{Y})\right)^2}{\sum_{i=1}^{n}(O_i-\overline{O})^2\sum_{i=1}^{n}(Y_i-\overline{Y})^2}, \tag{3}$$

$$E_{ns} = 1 - \frac{\sum_{i=1}^{n}(Y_i-O_i)^2}{\sum_{i=1}^{n}(Y_i-\overline{O})^2}, \tag{4}$$

Y_i is the calculated value, O_i is the measured value, $\overline{O}$ is the average value of the measured value, $\overline{Y}$ is the average value of the calculated value, and n is the number of data points.

3. Results

3.1. Rainfall Characteristics Analysis

During the 10 years from 2007 to 2016, there were 101 erosive rainfalls in the small Shixia watershed. In order to analyze the impact of rainfall on runoff and sediment yield, we had to eliminate the previous rainfall events as much as possible, so 45 precipitation events without rainfall in the previous two days were selected. The total rainfall of these 45 rainfall events was 1494.4 mm with an average rainfall of 33.2 mm, a maximum rainfall of 114.3 mm and a minimum rainfall of 9.8 mm. The average rainfall duration was 391.0 min, with a maximum rainfall duration of 1225.0 min, and a minimum rainfall duration of 20.0 min.

A total of 81 rain peaks appeared in 45 rainfall events (Table 1), and the maximum number of rain peaks that appeared in one rainfall event was five. There were 26 single-peak rainfall events, accounting for 41% of the total rainfall, and 19 multi-peak rainfall events, accounting for 59% of the total rainfall. In the multi-peak rainfall events, there were three rainfalls with a peak continuity of less than 1/90 min^{-1} and 16 rainfalls with a peak continuity more than 1/90 min^{-1}. Among the 81 rainfall peaks, the numbers that had a front position, middle position and rear position were 29, 30, and 22, respectively. Among the 45 rainfall events, the numbers that had a front position, middle position and rear position were 24, 13, and 8, respectively. It could be seen that the front-position rainfall in the research area was predominant and accounted for 53% of the total erosive storms.

Table 1. Statistics of the rainfall peak characteristics.

Rainfall Characteristics	Category	Number of Rainfall	Number of Peaks	Average Rainfall	Average Rainfall Duration	Average Rainfall Intensity
Peak Number	single-peak	26	-	23.67	297.23	12.75
	multi-peak	19	-	46.26	519.21	7.77
Peak position	front	24	29	22.75	235.25	12.66
	middle	13	30	48.81	532.85	7.25
	rare	8	22	39.25	627.50	10.12
Rain peak continuity	≤1/90 min^{-1}	3	-	30.17	998.33	1.72
	>1/90 min^{-1}	16	-	48.81	406.88	9.23

Figure 2 shows the frequency distribution of the rainfall peak value and peak width. The frequency distribution of the rainfall peak value in this region was in accordance with the Voiht peak shape function, and the peak value of rainfall was from 0.4 to 3.2 mm min^{-1}, of which 0.8–1.2 mm min^{-1} was most common, accounting for 33%. The frequency distribution of the rainfall peak width was in accordance with the extreme distribution. The peak width of rainfall was generally less than 60 min, of which 10–20 min was most common, accounting for 38%. The frequency distribution of a single rain peak width conformed to the extreme peak function. The duration of a single rain peak was generally below 40 min, and the majority were about 5–10 min, which accounted for 29%, followed by 10–15 min, which accounted for 28%.

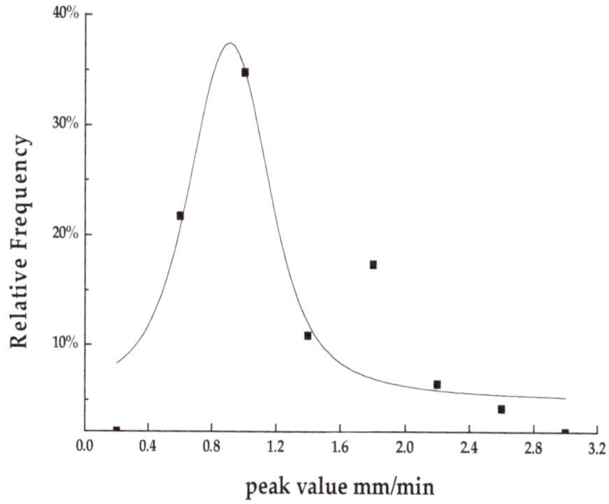

(a)

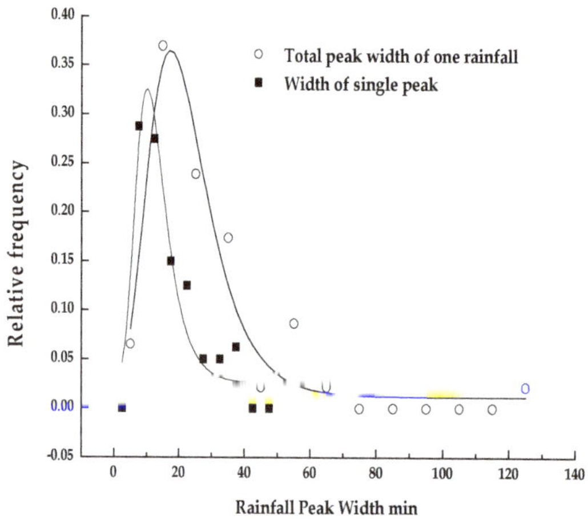

(b)

Figure 2. Frequency distribution of rainfall peak and peak width. (**a**) Frequency distribution of rainfall peak value, (**b**) frequency distribution of peak width.

3.2. *Effect of Rain Peak Morphology on Runoff and Sediment Yield*

3.2.1. Effect of Peak Number, Peak Position and Peak Continuity on Runoff and Sediment Yield

Taking the peak number of 46 rainfalls as independent variables, and runoff and sediment yield as dependent variables, one-way analysis of variance was carried out. The results are shown in Table 2. Table 2 shows that there were significant differences in runoff and sediment yield between single-peak and multi-peak rainfalls ($P < 0.05$). The average runoff of multi-peak rainfalls was 2.83 times that of single-peak rainfalls, and the average sediment yield of multi-peak rainfalls was 1.49 times that of single-peak. It could be seen that if the rainfall peaks reached two or more, the amount of soil and water erosion would increase significantly.

There was no obvious difference in runoff and sediment yield caused by rainfalls with peak continuity divided by 1/60 min^{-1}, and there was no significant difference in runoff caused by rainfalls with peak continuity divided by 1/90 min^{-1}. When the effect of multi-peak continuity on sediment yield was analyzed, the difference was found to be significant ($P < 0.05$). However, after excluding the effect of rainfall as a covariate, it was found that multi-peak continuity had a significant effect on sediment yield at the level of 0.1 ($P = 0.075$). With the increase in multi-peak continuity, the sediment yield showed an increasing trend. The average sediment yield of multi-peak rainfalls with continuity >1/90 min^{-1} was 2.95 times that of rainfalls with continuity ≤1/90 min^{-1}.

The peak position of 46 rainfalls were taken as independent variables, runoff and sediment yield were taken as dependent variables, rainfall amount were taken as covariates to exclude their influence, and covariance analysis was carried out. The results showed that there was no significant difference in soil erosion among front-position, middle-position, and rear-position rainfalls. However, the runoff formed by the front-position rainfall was significantly different from the runoff formed by the middle-position rainfall ($P < 0.05$), while the runoff and sediment yield formed by the rear-position rainfall was no significantly different from the runoff formed by front-position and middle-position rainfalls. The average runoff of the middle-position rainfall was 3.06 times that of the front-position rainfall, which indicated that the later the rain peak appeared, the more runoff that would be generated.

Table 2. Runoff and sediment yield under different peak numbers, peak positions, and peak continuities.

Rainfall Characteristics	Category	Average Sediment Yield (t km^{-2})			Total Sediment Yield (t km^{-2})	Average Runoff Yield (mm)			Total Runoff Yield (mm)
		Value	Standard Deviation	Figure		Value	Standard Deviation	Figure	
Peak number	single	625.17	522.19	a	16,254.42	6.45	5.21	a	167.93
	multi	969.87	552.58	b	18,427.53	20.05	11.62	b	380.95
Rain peak continuity	≤1/90 min^{-1}	363.01	229.42	a	1089.05	10.81	8.95	a	32.42
	>1/90 min^{-1}	1071.84	527.70	b	17,149.48	21.83	15.43	a	349.28
Peak position	front	625.31	526.98	a	16,350.00	6.55	5.28	a	157.43
	middle	1055.15	606.93	a	13,716.98	20.05	11.58	b	260.63
	rear	770.71	556.32	a	5708.69	16.39	13.02	ab	131.10

Note: Averages followed by the same letter in the same column are not significantly different, according to the Fisher LSD test at $P \leq 0.05$, P means significance test probability.

3.2.2. Effect of Peak Value and Peak Width on Runoff and Sediment Yield

Figure 3 shows the variation in runoff and sediment yield with the peak value. With an increase in the peak value, the average runoff and sediment yield increased gradually. The average runoff and sediment yield with a peak value >2 mm min^{-1} were 2.58 times and 2.09 times higher than those with a peak value in the range of 0.4–0.8 mm min^{-1}, respectively. Rainfall with a peak value from 0.8 to 1.2 mm min^{-1} produced the largest amount of runoff, which was up to 146.50 mm, accounting for 27% of runoff generated by all rainfalls. Although the average runoff of one rainfall within this peak range was the smallest, the number of rainfalls was the highest (15, see Figure 2), thus contributing greatly to the total runoff. Rainfalls with a peak value of 0.8–1.2 mm min^{-1} also produced the largest amount of sediment, which was up to 8825.33 t km^{-2}, accounting for 25% of the sediment generated by all rainfalls because of the high number of rainfalls.

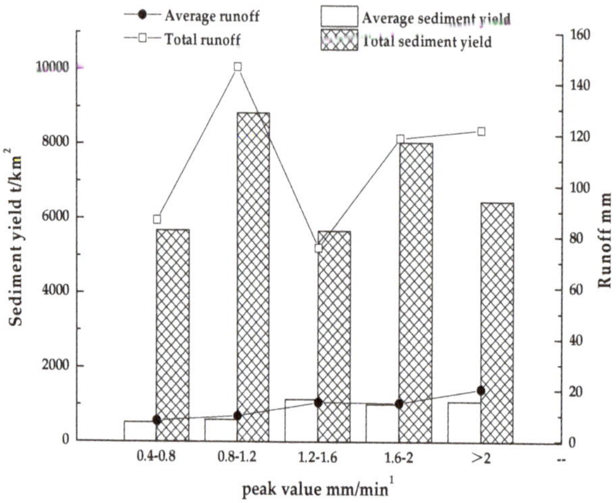

Figure 3. Runoff and sediment yield characteristics of different peak values of the rainfalls.

Figure 4 shows the runoff and sediment yield characteristics of rainfalls with different peak widths. With an increase in the peak width of the rainfall, the average runoff and the average sediment yield showed an increasing trend. Compared with rainfalls with a peak width <10 min, the runoff yield and sediment yield formed by rainfalls with a peak width >60 min increased by 7.16 times and 8.05 times, respectively. Rainfall with a peak width of 20–30 min produced the largest amount of runoff, which was up to 149.79 mm, accounting for 27% of runoff generated by all rainfalls. Rainfall with a peak width of 20–30 min not only produced high average runoff (13.62 mm) but also had high rainfall times (11, see Figure 2), thus contributing greatly to the total runoff. Rainfalls with a peak width of 10–20 min produced the largest amount of sediment, which was up to 9099.07 t km^{-2}, accounting for 26% of the sediment generated by all rainfalls. This was mainly due to the largest number of rainfalls within 10–20 min (17, see Figure 2), accounting for 38% of the total rainfalls.

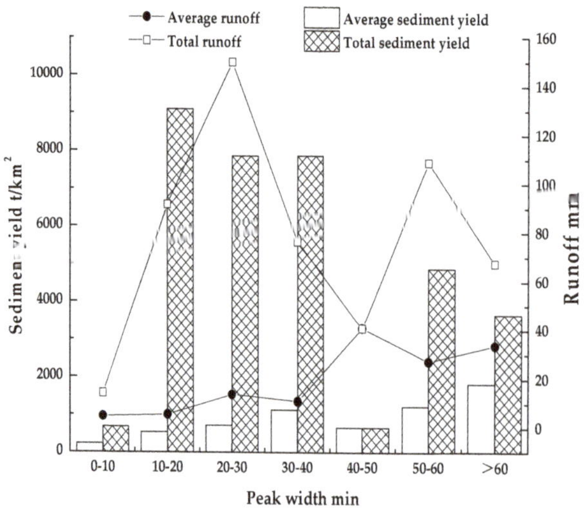

Figure 4. Runoff and sediment yield characteristics of rainfalls with different peak widths.

3.2.3. Comprehensive Effect of Rainfall Peak Morphology on Runoff and Sediment Yield

The results of the correlation analysis between runoff, sediment yield, and rainfall peak morphological indexes are shown in Table 3. The order of influence of rainfall peak morphology indexes on runoff was peak width (0.71) > peak number (0.69) > peak value (0.33) > peak position (0.05). The influence on sediment was peak width (0.62) > peak value (0.36) > peak number (0.36) > peak position (−0.09). The correlation of multi-peak continuity to runoff and sediment yield could not be compared with other indexes because of the small sample size. Table 4 shows that multi-peak continuity is not significantly correlated with runoff (0.12) and sediment yield (0.45). It is worth noting that the correlation coefficients of the peak width of one rainfall with the runoff and sediment yield were 0.71 and 0.62, respectively, which were much larger than that of the width of the highest peak with the runoff and sediment yield (0.36 and 0.49). Based on the statistical analysis of the influence of and the highest peak rainfall (*HPR*) on runoff and sediment yield, it was concluded that the influence of *TPR* on runoff and sediment yield was greater than that of *HPR*, which indicates that the explanatory power of runoff and sediment yield was weak when only the highest peak rainfall was considered.

Table 3. Correlation table between morphological indexes of rain peak with runoff and sediment yield.

Variable	Peak Number	Peak Continuity	Peak Position	Peak Value	Peak Width of a Rainfall	Highest Peak Width	*TPR*	*HPR*
Runoff	0.69 **	0.12	0.05	0.33 *	0.71 **	0.36 *	0.71 **	0.37 *
Sediment	0.36 *	0.45	−0.09	0.36 *	0.62 **	0.49 **	0.60 **	0.38 **

Note: * and ** mean significant differences at 0.05 and 0.01 level, respectively.

Table 4. Correlation of statistical characteristic indicators of rainfall with sediment yield and runoff.

Variable	Rainfall Amount	Average Rainfall Intensity	Rainfall Duration	I_5	I_{10}	I_{15}	I_{20}	I_{30}	I_{60}
Runoff	0.75 **	−0.08	0.30 *	0.33 *	0.31 *	0.40 **	0.47 **	0.44 **	0.69 **
Sediment	0.37 *	0.14	−0.15	0.36 *	0.28 **	0.38 **	0.44 **	0.48 **	0.60 **

Note: * and ** mean significant differences at the 0.05 and 0.01 level, respectively.

TPR was closely related to runoff and sediment yield, which might be due to the comprehensive consideration of the impact of multi-peaks, peak value and peak width, thus greatly enhancing the interpretation of runoff and sediment yield. However, *TPR* still failed to consider the peak continuity and peak position, which would weaken the explanatory power of runoff and sediment yield to a certain extent. Figure 5 shows the variation of runoff and sediment yield with the *TPR* of 45 rainfalls. From Figure 5, it can be seen that the trend of runoff and sediment yield was obviously consistent with *TPR*.

In this study, the annual variability coefficients of the runoff and sediment yield were calculated and their variations with *TPR* were plotted (Figure 6). The C_v of the runoff was from 0.49 to 1.24, and the C_v of the sediment yield was from 0.37 to 1.20. The C_v of the sediment yield was smaller than that of the runoff. The C_v value of the runoff and sediment was larger than that of *TPR* (0.41–0.97). With an increase in *TPR*, the C_v of the runoff and sediment yield also showed an increasing trend, which indicated that the runoff and sediment yield were more sensitive to heavy rainfall. That is, runoff and sediment yield caused by rainfall with high intensity and huge amount fluctuated more, which was more disadvantageous to watershed management, flood control, and disaster prevention management.

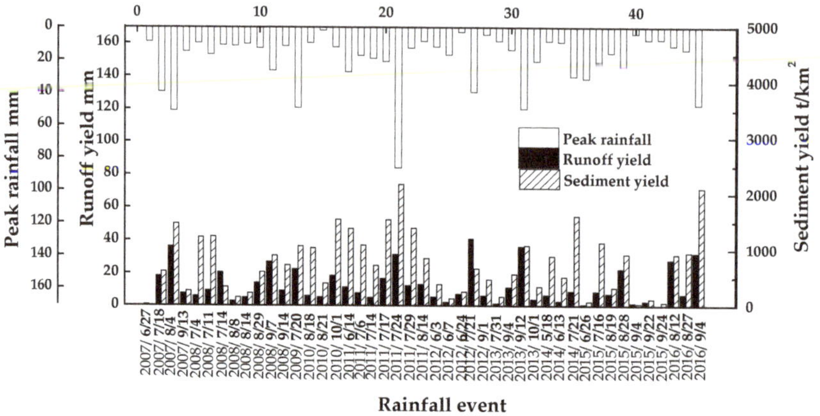

Figure 5. Variation of runoff and sediment with *TPR*.

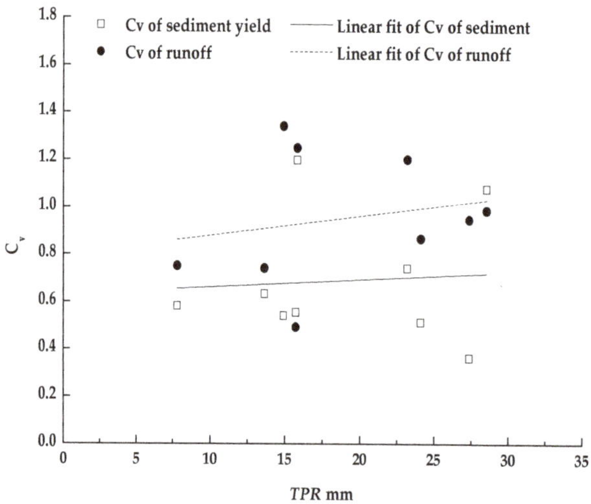

Figure 6. Change in the C_v of the annual runoff and sediment yield with the total peak rainfall amount of one rainfall (*TPR*).

3.3. Explanatory Power of Statistical Rainfall Indexes on Changes in Runoff and Sediment Yield

The correlation analysis between statistical rainfall indexes (rainfall amount, average rainfall intensity, rainfall duration, I_5, I_{10}, I_{15}, I_{20}, I_{30}, and I_{60}) and runoff and sediment yield were carried out. The results are shown in Table 4. Except for the average rainfall intensity and rainfall duration, there were significant correlations between the characteristic indexes of rainfall and sediment yield. The correlation between I_{60} and sediment yield was the largest (0.60). It can be seen from the foregoing that the duration of a single peak was generally less than 40 min (Figure 2), and its characteristics could be expressed by any index of the maximum rainfall of 5–30 min. The correlation between I_{60} and sediment yield was much greater than other statistical rainfall indexes, which indicated that the maximum rainfall of an intercept time of 60 min could accommodate more rainfall characteristics. However, it was not correct that the larger the interception time range was, the more the maximum rainfall in a given period is related to sediment yield because the correlation between rainfall amount and sediment yield was also low when the interception range was all time. It was probably related to the peak number and continuity in the rainfall process. I_{60} might reflect the comprehensive characteristics of multi-peak rainfall and their continuity to a certain extent, that was to say, the more continuous rain peaks occurred within one hour and the larger the peak value, the greater the sediment yield produced. The analysis of multi-peak continuity in Section 3.2.1 proved that when the peak interval is within a certain period of time (about 90 min), multi-peak continuity had a significant impact on sediment yield.

The trend of the relationship between runoff and maximum rainfall in a given period was consistent with sediment yield, except that the correlation between runoff and rainfall amount was also large. The statistical rainfall characteristic indexes with the strongest correlation with runoff were rainfall amount and I_{60} (Table 4), and the correlation coefficients were 0.75 and 0.69, respectively.

A comprehensive comparison between the indexes of rainfall peak morphology and statistical rainfall characteristic indexes showed that *TPR* and I_{60} were most closely related to runoff and sediment yield. Therefore, linear regression models were established with sediment yield (*S*) and runoff (*R*) as dependent variables, and *TPR* (*P*) and I_{60} as independent variables. The results are shown in Table 5. Five rainfall events were randomly sampled for cross-checking, and the relative error, correlation coefficient, and Nash–Sutcliffe efficiency coefficient of predicted values with measured values were obtained. The remaining rainfalls were used for relationship fitting. *TPR* could explain 62% of the runoff variety and 36% of the sediment yield variety, which was greater than explanatory power of I_{60} (51%, 37%). In general, *TPR* and I_{60} had more explanatory power for runoff than sediment yield. The cross-examination results of measured and simulated values of runoff yield model showed that the mean relative error of the two runoff models proposed in the study were 22.4% and 12.1%, the correlation coefficient was 0.91 and 0.94, and the Nash–Sutcliffe efficiency coefficient was 0.72 and 0.93, respectively. The models had good application precision. However, the mean relative error of the sediment yield model was larger, reaching more than 50%, and the Nash–Sutcliffe efficiency coefficient was less than 0, which indicated that its application precision was weak.

Table 5. Regression model of runoff and sediment yield.

Plot Code	Equations	R^2	*P*	Cross-Examination		
				MRE (%)	*R*	E_{ns}
1	$R = -0.25 + 0.64\ P$	0.62	<0.05	22.4	0.91	0.72
2	$R = -7.25 + 0.99\ I_{60}$	0.51	<0.05	12.1	0.94	0.93
3	$S = 376.68 + 20.02\ P$	0.36	<0.05	68.4	0.87	−1.86
4	$S = 34.91 + 36.93\ I_{60}$	0.37	<0.05	56.1	0.97	−0.19

Note: R^2 means coefficient of determination; *P* means significance test probability.

4. Discussion

Previous studies on how rainfall affects runoff and sediment yield have mostly focused on statistical descriptions of rainfall (rainfall, average rainfall intensity, duration of rainfall, and maximum rainfall in a given period), and maximum rainfall in a given time, with less than 30 min being the most common [19,20]. In this study, the correlation between rainfall and runoff was strong, but the correlation between rainfall and sediment yield was weak, and the average rainfall intensity had no significant correlation with runoff and sediment yield. This showed that the average rainfall intensity could not reflect the natural rainfall with instantaneous variation in rainfall intensity in the study of soil erosion. I_{30} could only reflect the characteristics of the highest rainfall peak, ignoring the situation of multiple peaks in the rainfall process. The results showed that there were significant differences in runoff and sediment yield between single-peak and multi-peak rainfalls. The continuity between rainfall peaks was rarely considered. In this paper, it was found that a distance of 90 min between the two rain peaks was a change point, and there were significant differences in soil loss caused by rainfall within and outside 90 min between the two peaks, which were concealed by correlation analysis.

The analysis of this study based on natural rainfall showed that the peak position had a significant impact on runoff, that is, the runoff caused by middle-position rainfalls was significantly larger than that caused by front-position rainfalls, while the peak position had no significant impact on sediment yield. Previous studies using simulated rainfall showed that more sediment could be generated by rear-position rainfall [18,21,22]. They believed that as the rain continued, the soil water content gradually increased, which led to the decline of infiltration capacity and an increase in runoff, or continuous high-intensity rainfall might lead to a change in soil structure. Flanagan [11] believed that when the soil was dry before a rainfall, a rear-position rainfall was more conducive to runoff generation, while if the soil was wet before a rainfall, a front-position rainfall could cause more runoff and soil loss. It was not found that the peak position had a significant effect on sediment yield in this study, which might be due to the different soil conditions under simulated rainfall and natural rainfall. Under artificial simulated rainfall conditions, the soil was relatively loose, which was not conducive to the formation of crusts [22]. Under natural conditions, the rear-position rainfall was favorable for the formation of a soil crust, that is, a storm with a lower initial rainfall intensity offered more opportunity for surface seal development, which was more resistant to erosion by surface flow [23]. On the contrary, the soil surface of the front-position rainfall under natural conditions was not protected by a crust, which was beneficial for sediment yield. Therefore, although the front-position rainfall produced less runoff than the middle- and rear-position rainfall, the sediment yield of the front-position rainfall was comparable to that of the middle- and rear-position rainfall. Frauenfeld and Truman [12] also found that the effect of rainfall patterns on sediment yield varied with soil type and texture.

Peak value and peak width had important effects on runoff and sediment yield, which could be expressed by *TPR*. The larger the peak value and width, the larger the *TPR*, and the larger the runoff and sediment yield. The results showed that the explanatory power of *TPR* for the runoff was 72%, with a higher correlation coefficient and Nash–Suttclife efficiency coefficient, and a relative error less than 30%, which indicated that the statistical model based on *TPR* had good application precision. The unexplained part might be due to: (1) This study based the rainfall events without anticipated rainfall within 2 days, which could not completely guarantee the consistency of soil water content before rain, and could not completely exclude the influence of soil water content. (2) The *TPR* could not accommodate the characteristics of rain peaks, such as peak position and continuity. (3) The nature of raindrops was not involved. (4) The interaction between rainfall and soil permeability.

The explanatory power of *TPR* on sediment yield was 36% with a low Nash–Sutcliffe efficiency coefficient, and a relative error more than 50%, which was similar to the research of Liu [24] on the application of the WEPP(water erosion prediction project) model to bare slope land in the black soil region of northeast China. This might be related to the type of soil and runoff process. Rainfall and runoff with different morphologies formed a complex response mechanism with soil particles, thus reducing the prediction accuracy of rainfall characteristic indicators on sediment yield. Further research is needed regarding how the interaction between rainfall and soil permeability affects runoff and how the interaction between rainfall, runoff and soil physical properties affects sediment yield.

5. Conclusions

(1) The order of influence of rainfall peak morphological indexes on runoff was peak width > peak number > peak value > peak position, and the order of the influence on sediment was peak width > peak value > peak number > peak position. Multi-peak continuity was not significantly correlated with runoff and sediment yield.

(2) Runoff and sediment yield increased significantly when the number of rainfall peaks reached two or more. The average interval time of 90 min between peaks of a multi-peak rainfall acted as the critical point, and there was a significant difference between the sediment with a peak continuity ≤1/90 min^{-1} and >1/90 min^{-1}. With an increase in peak continuity, the sediment yield showed an increasing trend. In this study area, a rainfall peak with an average interval of less than 90 min (continuity >1/90 min^{-1}) was most common, accounting for 84%.

(3) Analysis of variance showed peak position had no significant effect on soil erosion, but it had a significant effect on runoff. The runoff formed by middle-position rainfalls was significantly more than that of front-position rainfalls, which may be related to the formation of a soil crust.

(4) Peak number, peak value, and peak width had a correlation with runoff and sediment yield. With an increase in peak number, peak value and peak width, runoff and sediment yield also increased. As a comprehensive index, *TPR* could take into account all of the above factors and had the highest correlation with runoff and sediment yield.

(5) I_{30} could only reflect the rainfall characteristics of the highest peak because the rainfall peak width was generally below 40 min in the study area. However, I_{60} had a better explanatory power for runoff and sediment yield and could take more factors into account to a certain extent, such as multi-peaks and peak continuity.

(6) The regression equations established by *TPR* and I_{60} on runoff had good application precision. The prediction error was within 30%, and the correlation coefficient and the Nash–Sutcliffe efficiency coefficient were relatively high. Moreover, the interpretation of the *TPR* on the runoff was stronger than I_{60}. The precision of the regression equation established by *TPR* and I_{60} on the sediment yield was poor.

Author Contributions: Conceptualization, J.X.; Formal analysis, J.X.; Funding acquisition, F.W.; Investigation, M.L.; Methodology, J.X. and J.Z.; Project administration, J.Z. and F.W.; Resources, J.Z. and F.W.; Software, J.X. and M.L.; Supervision, J.Z.; Visualization, M.L.; Writing—original draft, J.X.; Writing—review & editing, J.X.

Funding: This research was funded by National Key Research and Development Program of China, grant number 2016YFC0501704.

Acknowledgments: We thank Shilei Wang and Xingle Yang (Engineer of soil and water conservation, Water and Soil Conservation Workstation in Miyun District, Beijing, China) for data collection and comments to earlier drafts of the manuscript.

Conflicts of Interest: The authors declare no conflict of interest.

Appendix A

Table A1. The data set of Statistical Rainfall Indexes.

Data	Rainfall Amount (mm)	Rainfall Duration (min)	Rainfall Intensity (mm/h)	I_5 (mm/h)	I_{10} (mm/h)	I_{15} (mm/h)	I_{20} (mm/h)	I_{30} (mm/h)	I_{40} (mm/h)	I_{50} (mm/h)	I_{60} (mm/h)
2007/6/27	26.6	560	2.85	43.20	36.60	35.20	29.10	23.00	18.60	16.44	15.90
2007/7/18	46.1	150	18.44	72.00	66.00	56.80	54.60	45.00	39.00	31.92	26.60
2007/8/4	66.5	375	10.64	109.20	96.60	77.20	72.60	66.00	52.65	45.00	39.60
2007/9/13	33.8	365	4.61	48.00	37.20	34.00	26.70	19.00	16.50	16.44	14.50
2008/7/4	17.5	430	2.44	84.00	57.60	44.80	34.05	23.00	17.25	13.80	11.50
2008/7/11	18.8	45	25.07	56.40	55.20	54.40	49.80	35.60	28.05	22.56	-
2008/7/14	51.9	1225	2.54	30.00	27.00	25.20	22.20	19.40	17.10	16.32	15.90
2008/8/8	16.8	750	1.34	42.00	36.00	30.00	24.00	16.80	15.00	14.16	12.50
2008/8/14	13.3	155	5.15	60.00	60.00	41.60	31.80	21.80	16.65	13.56	11.80
2008/8/29	21.8	1020	1.28	96.00	60.60	46.40	37.80	25.80	21.45	17.16	14.50
2008/9/7	34.5	240	9.00	108.00	100.20	85.60	66.30	48.80	42.45	35.28	30.40
2008/9/14	16.6	95	10.48	106.80	58.80	49.60	39.90	28.40	21.45	17.28	14.60
2009/7/20	54.8	220	15.66	127.20	117.60	92.00	77.40	54.00	40.80	32.76	27.70
2010/8/18	19.7	134	8.82	43.20	43.20	38.00	31.50	23.8	20.3	16.7	16.2
2010/8/21	43.4	1155	2.25	21.60	16.80	13.60	12.00	12.2	10.1	8.8	7.9
2010/10/1	38.5	618	3.74	30.00	30.00	29.60	28.80	27.0	24.2	22.1	20.6
2011/6/14	28.7	55	31.31	98.40	64.20	72.80	62.70	54.60	42.75	34.44	23.70
2011/7/6	30.1	320	5.64	60.00	43.80	36.20	36.90	30.20	26.40	21.84	15.70
2011/7/14	22.3	140	9.56	67.20	55.20	44.40	43.80	39.40	31.50	25.20	2[illegible].90
2011/7/17	32.6	225	8.69	67.20	54.60	49.20	42.90	33.80	28.65	30.12	2[illegible].70
2011/7/24	105.2	470	13.43	126.00	99.00	86.40	75.30	56.60	45.90	42.48	42.90
2011/7/29	20.1	185	6.52	76.80	58.20	50.80	43.20	32.20	25.65	21.12	18.00
2011/8/14	25	555	2.70	48.00	39.00	35.60	27.30	19.10	14.40	11.76	9.90
2012/6/3	12.40	20	37.20	100.80	71.40	48.80	36.90	24.80	18.45	14.76	12.40
2012/6/7	18.30	75	14.64	48.00	45.60	41.60	37.20	33.80	26.25	21.24	17.90
2012/6/24	29.90	825	2.17	32.40	23.40	19.60	16.20	12.40	11.10	9.60	9.10
2012/7/21	114.30	1040	6.59	69.60	51.00	47.20	43.50	34.00	31.80	28.32	24.60
2012/9/1	54.59	1095	2.99	27.60	26.40	18.80	14.40	11.60	11.40	11.04	10.50
2013/7/31	17.00	240	4.25	58.80	50.40	35.20	27.90	18.60	15.15	12.84	11.00
2013/9/4	62.50	625	6.00	42.00	34.20	34.80	33.60	30.40	25.95	21.72	18.50
2013/9/12	52.60	1215	2.60	138.00	99.00	82.00	72.30	30.00	55.50	49.20	41.00
2013/10/1	33.30	215	9.29	68.40	64.20	60.80	57.00	44.60	36.45	30.12	26.10
2014/5/18	9.8	40	14.7	33.60	33.00	25.20	18.90	6.7	14.7	11.76	9.3
2014/6/13	11.7	55	12.76	66.00	54.00	38.80	31.20	22.2	16.95	13.92	11.7
2014/7/21	30	85	21.18	108.00	84.00	73.2	67.2	57.6	45.75	36.72	30.6
2015/6/26	32.4	110	17.67	176.40	173.40	125.20	93.90	62.6	47.3	37.9	31.7
2015/7/16	26.8	160	10.05	151.20	79.80	54.40	41.40	28.6	33.5	28.0	23.3
2015/8/19	15.4	21	44.00	72.00	57.00	50.80	46.80	15.8	23.7	19.0	15.3
2015/8/28	28.7	110	15.65	146.40	103.20	83.60	72.00	54.0	41.1	33.5	28.0
2015/9/4	27.6	1060	1.56	24.00	21.00	18.00	15.00	10.0	11.7	10.9	9.8
2015/9/22	12.6	130	5.82	48.00	45.00	35.60	30.90	24.2	18.3	14.6	12.3
2015/9/24	11.0	80	8.25	60.00	44.40	32.80	25.50	19.0	14.4	11.6	10.3
2016/8/12	36.2	500	4.3	51.60	40.80	36.40	33.30	28.8	24.6	22.7	20.[illegible]
2016/8/27	15.1	35	25.9	108.00	81.60	57.20	43.65	30.0	22.5	18.0	15.[illegible]
2016/9/4	57.6	370	9.3	84.00	66.00	58.00	49.50	34.8	31.4	29.3	26.[illegible]

Table A2. The data set of morphological characteristics of rainfall peak.

Data	Peak Number	Peak Position	Peak Position Value [1]	Peak Width (min)	Highest Peak Width (min)	Peak Width of One Rainfall (min)	Highest Peak Value (mm)	Interval Time of Multi-Peak (min)	Total Interval Time of Multi-Peak (min)	Peak Continuity (min^{-1})	Peak Rainfall Amount (mm)	Highest Peak Rainfall Amount (mm)	Total Peak Rainfall Amount (mm)
2007/6/27	1	front	15	15	15	15	0.72	-	-	-	-	8.80	8.80
2007/7/18	2	rear	45	15/35	15	50	1.20	55	55.00	0.02	15/24.5	15.00	39.50
2007/8/4	4	rear	15	5/10/15/35	35	65	1.82	30/150/75	85	0.01	2/4.2/9.9/34.9	34.90	51.00
2007/9/13	3	front	5	5/5/15	5	25	0.80	75/25	50	0.02	2.3/4/8.5	4.00	14.80
2008/7/4	1	front	65	10	10	10	1.40	-	-	-	-	7.00	9.60
2008/7/11	1	front	5	20	20	20	0.94	-	-	-	-	16.60	16.60
2008/7/14	3	middle	345	5/15/5	15	25	0.50	380/45	212.5	0.00	2.5/6.3/2	6.30	10.80
2008/8/8	2	front	5	10/5	10	15	0.70	230	230	0.00	5.1/6.1	5.10	11.20
2008/8/14	1	front	0	10	10	10	1.00	-	-	-	-	10.00	10.00
2008/8/29	2	rear	125	5/10	10	15	1.60	850	850	0.00	2.6/10.1	10.10	12.70
2008/9/7	2	middle	95	5/15	15	20	1.80	30	30	0.03	5/21.4	21.40	26.40
2008/9/14	1	front	5	10	10	10	1.78	-	-	-	-	11.50	11.50
2009/7/20	2	middle	10	30/20	20	50	2.12	60	60	0.02	23.4/25.8	25.80	49.20
2010/8/18	1	middle	80	15	10	20	0.72	-	-	-	9.50	9.50	9.50
2010/8/21	1	middle	555	5	5	5	0.36	-	-	-	-	1.80	1.80
2010/10/1	1	middle	200	35	35	35	0.50	-	-	-	-	12.10	12.10
2011/6/14	1	front	15	30	30	30	1.64	-	0	-	-	27.30	27.30
2011/7/6	4	rear	0	10/10/5/5	10	30	1.00	215/5/5	75	0.01	6.6/7.3/3.3/2.5	7.30	17.10
2011/7/14	1	front	5	25	25	25	1.12	25	25	0.04	19.00	19.00	19.00
2011/7/17	2	front	5	25/10	25	35	1.12	40	40	0.03	16.1/6.7	16.10	21.00
2011/7/24	5	middle	265	35/25/5/5/50	25	120	2.10	15/10/10/5	10	0.10	10.9/11.2/27.2/2/32.7	11.20	86.20
2011/7/29	1	middle	70	15	15	15	1.28	-	-	-	-	12.70	12.70
2011/8/14	2	middle	220	15/5	15	20	0.80	70	70	0.01	8.5/3.3	8.50	8.50
2012/6/3	1	rear	150	10	10	10	1.68	-	-	-	-	11.90	11.90
2012/6/7	1	front	5	30	30	30	0.80	-	-	-	-	16.90	16.90
2012/6/24	1	rear	280	5	5	5	0.54	-	-	-	-	2.70	2.70
2012/7/21	5	middle	160	10/10/5/10/10	10	45	1.16	20/20/100/0	35	0.03	5.4/3.8/11.8/14.5/4.2	14.50	39.70
2012/9/1	1	rear	670	10	10	10	0.46	-	0.00	-	-	4.40	4.40
2013/7/31	1	front	20	10	10	10	0.98	-	0.00	-	-	8.40	8.40
2013/9/4	4	front	35	5/20/5/5	5	35	0.70	5/60/5	23.33333	0.04	2.5/11.2/3.5/2	3.50	13.70
2013/9/12	3	rear	25	15/20/20	20	55	2.30	60/25	42.5	0.02	9.7/24.1/16	24.10	49.80
2013/10/1	1	middle	60	25	25	25	1.14	-	-			20.80	20.80
2014/5/18	2	front	5	5/10	10	15	0.56	20	20	0.05	2.8/5.5	5.50	8.30
2014/6/13	1	front	0	10	10	10	1.10	-	-	-	-	9.00	9.00
2014/7/21	1	front	0	30	30	30	1.80	-	-	-	8.5/21.5	29.90	29.90
2015/6/26	1	front	0	15	15	15	2.94	-	-	-		31.30	31.30
2015/7/16	2	front	5	10/5	5	15	2.52	30	30	0.03	8.6/12.6	12.60	21.20
2015/8/19	1	front	5	20	20	20	1.20	-	-	-	-	15.60	15.60
2015/8/28	1	front	20	20	20	20	2.44	-	-	-	-	24.00	24.00
2015/9/4	1	front	335	5	5	5	0.40	-	-	-	-	3.80	3.80
2015/9/22	1	front	5	10	10	10	0.80	-	-	-	-	7.50	7.50
2015/9/24	1	front	0	10	10	10	1.00	-	-	-	-	7.40	7.40
2016/8/12	2	middle	50	5/20	20	25	0.86	-	-	-	-	11.20	11.20
2016/8/27	1	front	0	10	10	10	1.80	-	-	-	-	13.60	13.60
2016/9/4	3	middle	80	20/15/20	15	55	1.40	110/40	75	0.01	16.5/14/8.5	14.00	47.50

Note: [1] peak position value means distance from start time, min.

Table 3. The data set of runoff plots.

Date	Sediment (Plot 3)	Sediment (Plot 4)	Runoff (Plot 3)	Runoff (Plot 4)
2007/6/27	10.74	7.59	0.93	0.48
2007/7/18	641	596.55	25.04	11.54
2007/8/4	1509.81	1429.82	44.77	27.67
2007/9/13	286.31	245.33	7.9	7.34
2008/7/4	1149.43	1314.84	5.04	6.76
2008/7/11	1015.74	1469.81	9.09	9.27
2008/7/14	324.77	340.86	25.03	16.03
2008/8/8	102.09	198.3	2.79	3.02
2008/8/14	182.04	264.26	5.41	4.5
2008/8/29	539.84	672.23	13.5	14.4
2008/9/7	768.55	1045.17	29.62	24.22
2008/9/14	613.31	850.36	9	9.18
2009/7/20	1188.1	954.4	35.8	8.8
2010/8/18	1192.31	879.39	6.39	5.38
2010/8/21	460.21	329.86	6.39	4.13
2010/10/1	1716.51	1379.21	19.69	16.99
2011/6/14	1542.1	1234.93	13.45	8.95
2011/7/6	1118.74	1051.87	8.3	7.17
2011/7/14	739.71	707.67	5.93	4.19
2011/7/17	1527.13	1559.76	18.81	14.31
2011/7/24	2440.44	1904.11	38.7	23.86
2011/7/29	1353.62	1425.82	15.3	9
2011/8/14	755.24	944.5	17.06	8.96
2012/6/3	302.29	456.23	6.68	3.94
2012/6/7	175.01	80.45	2.65	2.09
2012/6/24	314.15	183.73	8.73	5.57
2012/7/21	578.96	740.84	28.27	53.47
2012/9/1	547.84	389.61	7.39	4.79
2013/7/31	222.19	104.92	1.77	1.1
2013/9/4	706.23	424.52	14.18	8.33
2013/9/12	1132.77	1028.77	37.62	34.92
2013/10/1	384.58	289.81	4.52	3.5
2014/5/18	689.31	1092.13	5.48	7.74
2014/6/13	630.61	394.5	2.77	2.99
2014/7/21	1738.23	1480.98	8.89	8.78
2015/6/26	89.56	62.94	0.75	0.87
2015/7/16	1330.75	949.2	8.91	8.57
2015/8/19	412	234.91	8.06	6.48
2015/8/28	688.07	1164.96	25.1	19.7
2015/9/4	51.51	18.67	2.35	1.11
2015/9/22	150.83	90.78	2.88	2.65
2015/9/24	82.48	31.16	0.19	0.1
2016/8/12	1161.85	663.57	28.8	[illegible]
2016/8/27	1245.27	651.07	6.78	7.23
2016/9/4	2534.04	1677.23	31.54	32.44

References

1. Fraser, A.I.; Harrod, T.R.; Haygarth, P.M. The effect of rainfall intensity on soil erosion and particulate phosphorus transfer from arable soils. *Water Sci. Technol.* **1999**, *39*, 41–45. [CrossRef]
2. Greer, J.D. Effect of excessive rate rainstorms on erosion. *J. Soil Water Conserv.* **1971**, *24*, 196–197.
3. Chaplot, V.; Le Bissonnais, Y. Runoff features for interrill erosion at different rainfall intensities, slope lengths and gradients in an agricultural loessial hillslope. *Soil Sci. Soc. Am. J.* **2003**, *67*, 844–851. [CrossRef]
4. Wang, W.Z.; Jiao, J.Y.; He, X.P. Study on rainfall erosivity in china. *J. Soil Water Conserv.* **1995**, *9*, 6–18.
5. Zhang, X.K.; Xu, J.H.; Lu, X.Q.; Deng, Y.J.; Gao, D.W. A study on the soil loss Equation in Heilongjiang Province. *Bull. Soil Water Conserv.* **1992**, *12*, 2–17.

6. Wischmeier, W.H.; Smith, D.D. *Predicting Rainfall Erosion Losses from Cropland East of the Rocky Mountains*, 1st ed.; United States Department of Agriculture: Washington, DC, USA, 1965; p. 282.
7. Bagarello, V.; D'Asaro, F. Estimating single storm erosion index. *Trans. Am. Soc. Agric. Eng.* **1994**, *37*, 785–791. [CrossRef]
8. Truman, C.C.; Bradford, J.M. Relationships between rainfall intensity and the interill soil loss–slope steepness ratio as affected by antecedent water content. *Soil Sci.* **1993**, *156*, 405–413. [CrossRef]
9. Assouline, S.; Ben-Hur, M. Effects of rainfall intensity and slope gradient on the dynamics of interrill erosion during soil surface sealing. *Catena* **2006**, *66*, 211–220. [CrossRef]
10. Arnáez, J.; Lasanta, T.; Ruiz-Flaño, P.; Ortigosa, L. Factors affecting runoff and erosion under simulated rainfall in Mediterranean vineyards. *Soil Tillage Res.* **2007**, *93*, 324–334. [CrossRef]
11. Flanagan, D.C.; Foster, G.R.; Moldenhauer, W.C. Storm pattern effect on infiltration, runoff, and erosion. *Trans. ASABE* **1988**, *31*, 414–420. [CrossRef]
12. Frauenfeld, B.; Truman, C. Variable rainfall intensity effects on runoff and interrill erosion from two coastal plain ultisols in Georgia. *Soil Sci.* **2004**, *169*, 143–154. [CrossRef]
13. Parsons, A.J.; Stone, P.M. Effects of intra-storm variations in rainfall intensity on interrill runoff and erosion. *Catena* **2006**, *67*, 68–78. [CrossRef]
14. Dunkerley, D. Effects of rainfall intensity fluctuations on infiltration and runoff: Rainfall simulation on dryland soils, Fowlers Gap, Australia. *Hydrol. Process.* **2012**, *26*, 2211–2224. [CrossRef]
15. Qin, W.; Zuo, C.Q.; Yan, Q.H.; Wang, Z.Y.; Du, P.F.; Yan, N. Regularity of individual rainfall soil erosion in bare slope land of red soil. *Trans. Chin. Soc. Agric. Eng.* **2015**, *31*, 124–132.
16. Hudson, N.W. *Soil Conservation*, 1st ed.; B.T. Batsford Ltd.: London, UK, 1976.
17. Huff, F.A. Time distribution of rainfall in heavy storms. *Water Resour. Res.* **1967**, *3*, 1007–1019. [CrossRef]
18. Wang, W.T.; Yin, S.Q.; Xie, Y.; Liu, B.Y.; Liu, Y.N. Effects of four storm patterns on soil loss from five soils under natural rainfall. *Catena* **2016**, *141*, 56–65. [CrossRef]
19. Wischmeier, W.C.; Smith, D.D. Rainfall energy and its relationship to soil loss. *Trans. Am. Geophys. Union* **1958**, *39*, 285–291. [CrossRef]
20. Kinnell, P.I.A. Runoff ratio as a factor in the empirical modelling of soil erosion by individual rainstorms. *Aust. J. Soil Res.* **1997**, *35*, 1–13. [CrossRef]
21. An, J.; Zheng, F.L.; Han, Y. Effects of rainstorm patterns on runoff and sediment yield processes. *Soil Sci.* **2014**, *179*, 293–303. [CrossRef]
22. De Lima, J.L.M.P.; Souza, C.S.; Singh, V.P. Granulometric characterization of sediments transported by surface runoff generated by moving storms. *Nonlinear Proc. Geophys.* **2008**, *15*, 999–1011. [CrossRef]
23. Romkens, M.J.M.; Helming, K.; Prasad, S.N. Soil erosion under different rainfall intensities, surface roughness, and soil water regimes. *Catena* **2001**, *46*, 103–123. [CrossRef]
24. Liu, Y.L.; Zheng, F.L.; Wang, B.; Wang, Y.X.; Xie, Y.J.; Fan, H. Assessment of WEPP model applicability in black soil zone of northeast China: A case study of slope gradient and soil and water conservation measures. *Bull. Soil Water Conserv.* **2010**, *30*, 139–145.

water

MDPI

Article

Design of a Pressurized Rainfall Simulator for Evaluating Performance of Erosion Control Practices

Matthew D. Ricks [1,2,*], Matthew A. Horne [3], Brian Faulkner [1], Wesley C. Zech [4], Xing Fang [1], Wesley N. Donald [1] and Michael A. Perez [1]

1 Department of Civil Engineering, Auburn University, Auburn, AL 36849-5337, USA; baf0015@auburn.edu (B.F.); xing.fang@auburn.edu (X.F.); donalwn@auburn.edu (W.N.D.); map0032@auburn.edu (M.A.P.)
2 Liberty Utilities, 2300 Victory Drive, Columbus, GA 31902, USA
3 HDR, 25 W. Cedar Street, Suite 200, Pensacola, FL 32502, USA; Matthew.Horne@hdrinc.com
4 Department of Civil, Construction, and Environmental Engineering, The University of Alabama at Birmingham, Birmingham, AL 35294-4440, USA; zechwes@uab.edu
* Correspondence: mzr0062@auburn.edu; Tel.: +1-706-566-9798

Received: 16 October 2019; Accepted: 10 November 2019; Published: 14 November 2019

check for updates

Abstract: Construction site erosion and resulting sedimentation constitutes one of the greatest non-point source pollution threats to our nation's waterways. Erosion control practices are important aspects of any construction project due to their ability to limit the process of erosion. Testing erosion control practices under simulated rainfall representative of conditions experienced on construction sites is important to better understand their erosion reduction capabilities. Full-scale testing using simulated rainfall has been shown to provide controllable and repeatable results, in comparison to field-testing under natural conditions. Therefore, the focus of this study was to design, construct, and calibrate a pressurized rainfall simulator testing apparatus capable of accurately and repeatedly simulating rainfall intensities of 50.8, 101.6, and 152.4 mm/hr (2.0, 4.0, and 6.0 in/hr) for 20-min intervals. The developed testing apparatus consisted of a 12 m (40 ft) long by 2.4 m (8.0 ft) earthen slope at a 3H:1V slope. Ten sprinkler risers at a height of 4.27 m (14 ft) were installed around the perimeter of the slope to create a uniform distribution of rainfall. Data collection procedures consisted of collecting and analyzing rainfall depth, drop size distributions, and sediment concentrations. The optimum location for each sprinkler riser, as well as the most accurate nozzle configuration, were determined through test procedures developed for this study. Through calibration testing, the simulator was found to produce accurate rainfall intensities with relative errors of 1.17–4.00% of the target intensities. Uniformity of rainfall distribution ranged from 85.7 to 87.5%. Average drop sizes were determined to be between 2.35 and 2.58 mm (0.093 to 0.102 in.).

Keywords: erosion control; full-scale testing; runoff; simulated rainfall; water quality

1. Introduction

Construction projects typically create large areas of exposed soil due to clearing, grubbing, and land grading activities. Lack of vegetative cover leaves these areas susceptible to erosion during rain events. In the U.S. alone, it is estimated that as much as 73 million metric tons (80 million tons) of sediment is eroded from construction sites each year [1]. Highly concentrated sediment-laden stormwater runoff degrades existing ecosystems and water quality through the process of sedimentation and by increasing turbidity, making it difficult for aquatic organisms to survive. These concerns led the U.S. Congress to include sediment discharge into the National Pollutant Discharge Elimination System (NPDES) permit program in 1990 under the Clean Water Act. Regulations under NPDES require sediment pollution generated by construction activities be controlled on-site by the site operator [2].

Several factors contribute to the erosive potential of a particular project site, including soil properties, topography, local climate (i.e., rainfall intensity, frequency, and duration), and vegetative cover. While there are many different types of erosion (e.g., interrill, rill, gully, channel, etc.) and subsequent control practices and products, the focus of this paper is on the testing and evaluation of erosion control practices and products used on earthen slopes to minimize interrill and rill erosion.

With the increasing usage of erosion control practices and products, it is important for researchers, practitioners, contractors, inspectors, and regulatory agencies to understand their in-field performance along with suitable applications. Small-scale testing has been conducted in the past to accomplish these objectives; however, it does not adequately represent conditions that practices and products would experience in the field [3]. To effectively recreate field-like scenarios, full-scale testing on a field-scale plot must be performed. To date, the most representative and controllable method to accomplish this has been through the use of rainfall simulators [4–7]. The purpose of this study was to design and calibrate a test apparatus that is capable of evaluating the effectiveness of erosion control practices and products, and enables the researchers to replicate these results.

1.1. Rainfall Simulators

Rainfall simulation has long been used to study the effects of rainfall-induced erosion [8–12]. The need for rainfall simulators arose when researchers determined that simulated rainfall provided more uniform control over experiments in comparison to natural rainfall. While natural rainfall is most desirable for testing of erosion control practices, simulated rainfall allows for expedited data collection and reproducible testing [13–15].

The earliest rainfall simulators used drop forming mechanisms (i.e., hypodermic needles and string) to form droplets [6]. Unpressurized systems need to release raindrops from heights of up to 9.1 m (30 ft) to ensure they reach terminal velocity, representative of natural rainfall. Furthermore, these systems are highly susceptible to environmental conditions (i.e., wind), leading to these type of simulators being employed almost exclusively in enclosed laboratory settings.

Beginning in the mid-20th century, pressurized rainfall simulation systems became more desirable to conduct large-scale, outdoor experiments [7,9,14,16,17]. Pressurized rainfall simulators rely on nozzles or sprinkler heads to produce rain-like droplets. With a pressurized system, raindrops have the ability to reach terminal velocity quickly, thereby allowing for shorter, more portable simulators. Furthermore, pressurized rainfall simulators provide some resistance to environmental conditions, allowing researchers to conduct evaluations outdoors.

Moore et al. [14] designed, what is referred to as a Kentucky rainfall simulator, using the following four criteria to generate conditions similar to natural rainfall: (1) uniform distribution, (2) rainfall intensities, (3) drop size distributions, and (4) raindrop velocities that create kinetic energy. Furthermore, a plot size large enough to effectively simulate field-like conditions is required. In addition to the above criteria, Meyer [18] identified five supplementary design criteria that must be satisfied to adequately simulate natural rainfall: (1) intensities similar to storms producing medium to high rates of runoff and erosion, (2) near continuous rainfall application, (3) near vertical impact of most drops, (4) satisfactory performance in windy conditions, and (5) portability of the system.

1.2. Previous Erosion Studies Using Simulated Rainfall

Pressurized rainfall simulators (Table 1) differ between studies due to varying research objectives and plot sizes. Shoemaker et al. [19] developed a laboratory-scale rainfall simulator to conduct studies on the effectiveness of anionic polyacrylamide as an erosion control measure. The simulator consisted of a single solenoid operated nozzle. The nozzle was installed at a height of 3.05 m (10 ft) and used a pressure regulator to control flow. Two 3H:1V sloped plots, each with a surface area of 0.74 m^2 (8.0 ft^2), were constructed and placed under the simulator. The nozzle was capable of producing a rainfall intensity of 11.2 cm/hr (4.4 in./hr). Tests consisted of four, 15-min rainfall events separated by 15-min intervals of no rainfall to allow for data collection. Using Christiansen's Uniformity Coefficient (CUC),

Shoemaker et al. [19] calculated an average uniformity of rainfall distribution of 83 to 87% on the test plots.

Table 1. Summary of rainfall simulators and testing.

Study	Drop Size Distribution, mm (in.)	Uniformity	Simulator Height, m (ft)	Rainfall Intensity, mm/hr (in./hr)	Plot Sizes, m^2 (ft^2)	Slopes, %
ASTM [a] D6459-15 [22]	Less than 10% > 6 (0.24) Less than 10% < 1 (0.04)	>80%	4.27 (14)	50.8, 101.6, 152.4 (2, 4, 6)	29.7 (320)	33
Moore et al. [14]	D50 = 2.25 (0.089)	80.2 to 83.7	3 (9.84)	3.5 to 185 (0.138 to 7.28)	4.5 (48.4) or 99 (1065)	
Shoemaker et al. [19]		83 to 87	3.05 (10)	111.8 (4.4)	0.74 (8)	33
Kim et al. [20]			2.44 (8)	71.12 to 83.82 (2.8 to 3.3)	8 (86)	29 to 30
McLaughlin and Brown [9], Miller [21]	2.25 to 2.5 (0.089 to 0.098)	85.7 to 93.2	3.96 (13)	33 and 66 (1.3 and 2.6)	2 (21.8)	10 and 20

Note: [a] ASTM International.

Kim et al. [20] conducted a study examining the effectiveness of flocculant treatments on steep vegetable fields in South Korea. Six test plots were constructed on slopes ranging from 29% to 30% with surface areas of 2.4 m^2 (26 ft^2). Kim et al. [20] constructed a rainfall simulator with steel angle iron and sprinklers set at a height of 2.4 m (8.0 ft). The simulator was capable of generating rainfall intensities from 70 to 85 mm/hr (2.8 to 3.3 in./hr).

McLaughlin and Brown [9] conducted a rainfall simulation study with the objective of determining if application of flocculant to mulches provided erosion control improvements. For this study, 1 m (3.3 ft) wide by 2 m (6.6 ft) long test plots were constructed on slopes of 10 and 20%. A rainfall simulator based on a similar design to that of Miller's [21] was constructed for this experiment. A 1/2HH-SS50WSQ Fulljet nozzle (Spraying Systems Co.®, Wheaton, IL, USA) was installed 3.96 m (13.0 ft) above the test plots to produce rain drops. The nozzle was set at a pressure of 34 kPa (5.0 psi) and produced droplet sizes similar to natural rainfall. During tests, the simulator produced constant rainfall intensity of 68 mm/hr (2.6 in./hr). The intensity was reduced to a rate of 33 mm/hr (1.3 in./hr) by programming a solenoid valve to cycle off-and-on in 10 s intervals. Tests were performed until 5 min after runoff was observed from the test plots.

2. Current Standard Test Methods and Installment Procedures

ASTM D6459-15 [22] is the ASTM International standard test method for determining the performance of rolled erosion control product (RECP) using rainfall simulation. This standard test method is used to quantify rainfall-induced erosion of hillslopes under the protection of RECPs [22]. The test determines the soil erodibility factor, K, of the soil used and the cover management factor, C, of a RECP tested. The data analysis allows for the comparison of C values of different RECPs to understand their relative performance for controlling erosion. To determine K and C from the Revised Universal Soil Loss Equation (RUSLE), the rainfall-runoff erosivity factor, R, must first be determined, which is calculated from the erosion index (EI) using Equation (1) [23]:

$$R = \frac{1}{n}\sum_{j=1}^{n}[\sum_{k=1}^{m} EI_k] \tag{1}$$

where EI_k is the erosion index for the rainfall event k, m is the total number of the rainfall events in a year, and n is the number of years used to obtain average R in hundreds of ft·tonf·in./(acre·hr·yr). EI is the product of total storm energy (E) multiplied by the maximum 30-min intensity (I_{30}) for a given storm event, where E is in hundreds ft·tonf/acre and I_{30} is in in./hr [23]. E is the total kinetic energy of all raindrops of the storm and directly related to the rainfall intensity. Since R is the average erosivity potential from a known set of storm events over a known period of time, each storm within that time period must be individually analyzed using Equation (2) to determine erosivity for each storm event:

$$EI = (E)I_{30} = \left(\sum_{r=1}^{p} e_r \Delta V_r\right) I_{30}\left(10^{-2}\right) \tag{2}$$

where e_r is the rainfall energy per unit depth of rainfall per unit area in ft·tonf/(acre·in.), ΔV_r is the depth of rainfall (in.) for the rth increment of the storm hyetograph which is divided into p parts, each with essentially constant rainfall intensity. For natural rainfall events, each raindrop reaches its terminal velocity when reaching the ground, and e_r can be calculated as a function of rainfall intensity i_r (in./hr) using Equation (3) in the United States [23]:

$$e_r = 1099[1 - 0.72exp(-1.27i_r)] \; and \; i_r = \Delta V_r / \Delta t_r \tag{3}$$

For large-scale testing in ASTM D6459-15, evaluations for each RECP are repeated three times, therefore, an annual R cannot be developed, only an average EI and C for three tests are determined for comparing the relative performance of different RECPs. ASTM D6459-15 suggests using Equation (4) to compute EI [22]:

$$EI = I \times 1099 \times \left[1 - 0.72e^{(-1.27I)}\right] \tag{4}$$

ASTM D6459-15 does not provide any specific details to apply Equation (4) but refers to the U.S. Department of Agriculture (USDA) Agriculture Handbook 703 [23]. Equation (4) as described by ASTM D6459-15 misrepresents the original Equations (2) and (3) since there are two Is in Equation (4) that represent two different intensities: the first I represents I_{30} and the second I is intended to represent i_r for the rth increment of the storm hyetograph. Also, when using the rainfall simulator (not natural rainfall), each rainfall drop may not reach its terminal velocity and the unit rainfall energy, e_r, cannot be calculated using Equations (3) or (4) directly. In this study, we will discuss the use of the original method in the USDA Agriculture Handbook 703 [23] to directly compute the erosion index for the large-scale test.

Following ASTM D6459-15, the rainfall simulator includes the use of sprinkler heads, sprinkler risers, pressure gauges, and valves. The ASTM design consists of nine sprinkler risers spaced evenly around the test plot. Raindrop sizes should vary from 1.0 to 6.0 mm (0.04 to 0.25 in.). Furthermore, the risers should be constructed to generate a minimum raindrop fall height of 4.3 m (14 ft). To conduct large-scale testing, a 12 m (40 ft) long by 2.4 m (8.0 ft) wide test plot must be constructed on a 3H:1V slope. The soil veneer used for testing should be placed in two, 15 cm (6.0 in.) lifts and must consist of either a loam, sand, or clay soil. The drop size distribution for a specific intensity is determined using the flour pan method [21,25]. Specified rainfall intensities are 50.8, 101.6, and 152.4 mm/hr (2.0, 4.0, and 6.0 in./hr). The test consists of three, 20-min intervals of increasing rainfall intensity for a total of 60 min.

The ASTM standard requires apparatus calibration to ensure experimental values for uniformity of rainfall distribution, rainfall intensity, and drop size distribution are similar to natural rainfall. A calibration test consists of running the simulator at a specific intensity for 15 min. A collection of 20 rain gauges should be spaced throughout the test plot to collect rainfall data. The recorded rainfall depth in each rain gauge is analyzed to determine the experimental values for rainfall uniformity and intensity.

3. Design and Construction of the Auburn Rainfall Simulator

The purpose of this manuscript is to document the design and construction of a pressurized rainfall simulator with the aim of developing a portable, calibrated simulator capable of producing replicable, simulated rainfall events. The development of this rainfall simulator is in collaboration with the Alabama Department of Transportation to provide the capability to evaluate erosion control practices and products based upon performance under simulated rainfall. This simulator will also provide the industry with much need additional testing options, as there is currently only one simulator within the U.S. that is certified to ASTM D6459 and American Association of State Highway and Transportation Officials (AASHTO) testing requirements. In this study, a rainfall simulator and full-scale test plot were constructed at the Auburn University-Erosion and Sediment Control Test Facility (AU-ESCTF) in Opelika, Alabama. The rainfall simulator design was based on specifications listed in ASTM D6459-15 [22].

3.1. Sprinkler Head Design

As specified in ASTM D6459-15 [22], the targeted rainfall intensities for the rainfall simulator was designed for 20-min intervals of 50.8, 101.6, and 152.4 mm/hr (2.0, 4.0, and 6.0 in./hr), respectively. The rainfall simulator apparatus incorporates solenoid valves to instantaneously alter flow rates by turning on sprinklers to achieve the required intensity over the 60-min experiment.

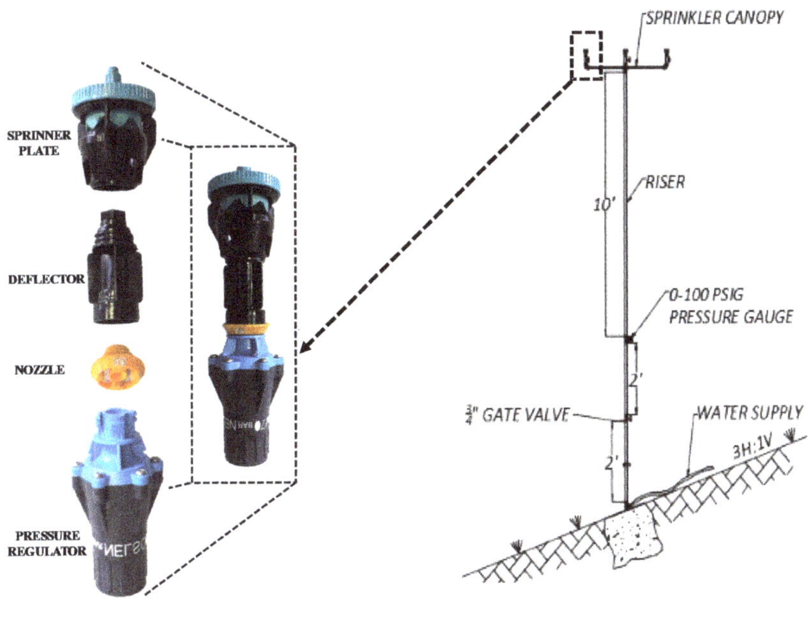

(a)

Figure 1. *Cont.*

(b)

Figure 1. Pressurized rainfall simulator layout and components [26]. (**a**): canopy, riser, and anchor detail; (**b**): rain gauge layout constructed.

After reviewing several commercially available pressurized sprinkler heads, Nelson Irrigation (Walla Walla, Washington, USA) PC-S3000 sprinkler heads Figure 1a were selected in lieu of nozzles specified in ASTM D6459-15, as the specified nozzles are no longer commercially available. The PC-S3000 sprinkler heads were selected in part due to: their ability to operate at pressures as low as 41.4 kPa (6.0 psi); apply water in a 190° arc; the ability of the equipped spinner plates to shear apart flow in the sprinkler head to generate rain-like droplets; and their capability in interchangeable nozzles that allow for various flow rates. The PC-S3000 use nozzles to control the flow rate through each sprinkler head. At any given pressure, each nozzle allows a specific flow rate through the sprinkler, depending on its size. Furthermore, the sprinkler heads can be equipped with pressure regulators to ensure uniform pressure and thereby a constant flow rate. Manufacturer specifications for each nozzle size were used to determine the appropriate nozzle sizes for this study.

Several combinations of selected nozzles (Table 2) were used to achieve the variable intensities required to simulate the rainfall event. Initial testing was conducted for the selected nozzle sizes to determine the optimal spacing of sprinkler heads in relation to the rainfall plot gauges. This testing was conducted for each sprinkler considered to determine the optimal distance from a riser to the predetermined rain gauge layout as shown above. One example of the test data is shown in Figure 2, below. The 15-min duration tests were repeated many times for each sprinkler to generate a data series of rainfall depths at nine rainfall gauges (different distances to the sprinkler) in order to develop the box plots on Figure 2. At three rainfall gauges (1.22, 1.72, and 3.66 m) measured rainfall depths did not have much variation so no box plots are shown for them on Figure 2.

Table 2. Nozzle combinations.

20-min Test Interval	Number and Type of Nozzles Used	Total Flow, L/min. (gpm)	Theoretical Flow Requirement, L/min (gpm)	Theoretical vs. Total Flow (%)
1	10–#21 [a]	18.4 (69.65)	6.65 (25.17)	36.1
2	15–#21 5–#18 [b]	34.4 (130.22)	13.30 (50.35)	38.7
3	21–#21 9–#18	50.9 (192.68)	19.95 (75.52)	39.2

Note: [a] #21–Turquoise yellow nozzle, flow rate at 41.4 kPa (6 psi) = 6.96 L/min. (1.84 gpm); [b] #18–Gray nozzle, flow rate at 41.4 kPa (6 psi) = 5.14 L/min (1.36 gpm).

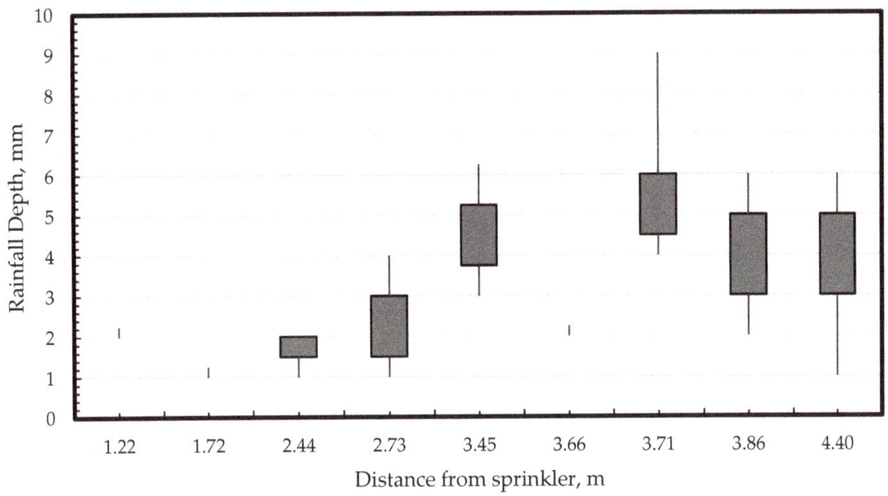

Figure 2. Rainfall depth relative to distance from Nelson PC-S3000 sprinkler head with #21 nozzle at 41.4 kPa (6 psi) for 15-min duration tests.

Based on the testing conducted, the following distances were selected for the riser spacing: 1.52 m (5.0 ft) from plot edge and 3.05 m (10.0 ft) center to center as shown in Figure 3. As detailed in ASTM 6459-15 [22], nine risers are specified for the test plot. For this study, an additional riser was installed at the top of the test slope as shown below to provide for a more uniform delivery of simulated rainfall across the plot. As shown in Figure 3, the 190° application allows for a large portion of the rainfall to be applied outside the plot area. The ratio of the plot area of application to the overall sprinkler application area is approximately 36%. As shown in Table 2, the percentage for the theoretical flow requirement (rainfall depth x plot area) versus the total flow produced (total output for all nozzles) is fairly consistent throughout all three rainfall intensities.

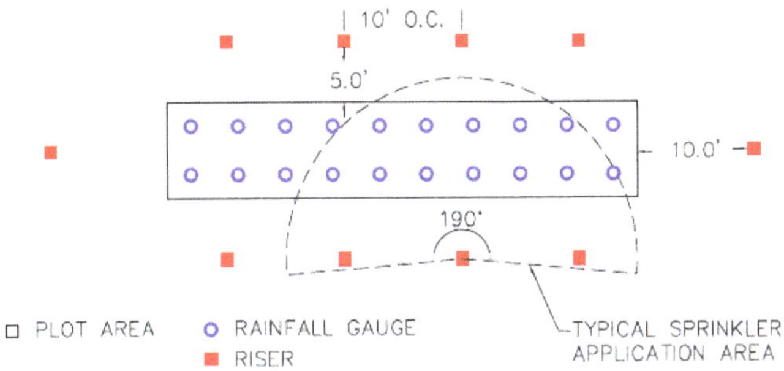

Figure 3. Detailed plan view of riser and rainfall gauge locations.

3.2. Sprinkler Canopy Design

To effectively distribute water over the test plot, a rain canopy, Figure 1b, with three sprinkler heads was designed for each of the ten risers. The canopy was designed to allow for each sprinkler head to be individually operated to achieve flow rates displayed in Table 2. The Nelson Irrigation sprinkler heads, Figure 1a, spray directly out and downward, and the height of the risers was set at 4.3 m (14 ft) to satisfy ASTM D6459 standard for fall height and terminal velocity.

The canopy and all components were constructed of 19 mm (0.75 in.) diameter galvanized steel pipe to provide structural stability as well as to resist corrosion. The canopy connected to the supporting riser through a galvanized steel pipe cross in the center of the canopy. Solenoid valves were installed upstream of each sprinkler head to allow for individual and automated operation. The use of solenoid valves represents a change from the manually operated ball valves in ASTM D6459-15.

3.3. Sprinkler Riser Design and Water Supply

A reducer tee was installed to allow for the attachment of a 0 to 690 kPa (0 to 100 psi) pressure gauge. This gauge was used to verify that adequate pressure was generated by the water delivery system. The riser is supported by a concrete footing as shown in Figure 1b. The risers were also anchored to a 10.2 cm by 10.2 cm (4 in. by 4 in.) nominal pressure treated post to ensure the riser remains plumb and level. A 7.62 cm (3 in.) high pressure water pump, capable of producing 80.2 m (263 ft) of pump head, was used to supply and pressurize water to the rainfall simulator.

3.4. Wind Screen Design

To minimize the impact of cross winds on rainfall simulation experiments, a series of wind screens, suspension cables, and support posts were designed with the goal of reducing wind speeds on the plot to at most 1.6 km/hr (1.0 mi/hr). To support the screens, six, 12 cm by 12 cm, 6.0 m (6.0 in. by 6.0 in., 20 ft) nominal lumber posts were installed around the perimeter of the test plot, Figure 1b.

3.5. Electrical Systems Design

Simulation of variable intensity rainfall was accomplished by installing solenoid valves on the sprinkler canopy. The valves on each canopy were wired via direct burial irrigation cable to a custom designed electrical control box. The electrical control box consisted of a series of terminal blocks and was designed with three switches to provide control over which valves were active during testing. Two 12-V batteries were wired in parallel to the control box to power the entire valve system. Utilization of electronically controlled valves is an improvement over the current standard [22] which has been utilized in other rainfall simulators [27].

3.6. Methods and Procedures

Initially, the rainfall simulator apparatus was calibrated to determine the experimental values for rainfall intensity and uniformity. This process was critical in proving accurate and repeatable simulated conditions similar to natural rainfall (i.e., uniformity, drop size, and terminal velocity).

According to ASTM D6459-15 [22], twenty rainfall gauges are required when measuring and calibrating rainfall intensity and distribution. For this study, an additional nine rainfall gauges were installed along the center of the test plot as shown in Figure 1b.

For each target rainfall intensity, a calibration test was performed for a duration of 15 min. At the end of the test, the rainfall depth in each of the 29 gauges was measured and recorded in centimeters. The recorded values for rainfall depth were then used to calculate CUC using Equation (5), and average rainfall intensity:

$$CUC = 100\left[1.0 - \frac{\sum\left(\left|D_i - D_{avg}\right|\right)}{n \times D_{avg}}\right] \tag{5}$$

where CUC = Christiansen's Uniformity Coefficient used to express uniformity of rainfall (%), D_i = depth of rainfall in the ith gauge (cm), D_{avg} = average rainfall depth in all gauges (cm), and n = number of gauges.

Using Equation (6), experimental rainfall intensities on the test plot were computed and compared to the targeted rainfall intensities for determining the relative errors:

$$i = 60\left[\sum_{j=1}^{J} \frac{D_j}{Jt}\right] \tag{6}$$

where i = rainfall intensity (cm/hr), D_j = depth of rainfall (cm), J = number of rain gauges, and t = test duration (min).

Once the uniformity of rainfall was at least 80% for each target intensity, the raindrop size distribution for each intensity was measured using the flour pan method [24,25]. For each intensity, pans were filled with sifted flour and exposed to rainfall for 2.0 to 4.0 s. Raindrops impacting the flour created small pellets that were then sifted, baked, and separated using sieves. The pellets on each sieve were then weighed and counted. This process was repeated three times (at the top, middle, and bottom part of the slope) for each test intensity. Each of the four steps for the flour pan method are depicted in Figure 4. Using this information, the average raindrop diameter for each sieve was then calculated using Equation (7) [25]:

$$D_r = \sqrt[3]{\left(\frac{6}{\pi}\right) M\, m_R} \tag{7}$$

where D_r = average raindrop diameter (mm), M = average pellet mass (mg), which is the total mass divided by the number of pellets in each sieve for all three repetitions, and m_R is the ratio of the mass of the raindrop to the mass of the pellet and determined using the flour-calibration figure developed by Laws and Parsons [25]. At the same time, the percent of the mass of the raindrops for each sieve can be determined with respect to total mass of all raindrops measured. The calculated results are shown in Figure 5, below.

Figure 4. Drop size distribution testing procedures. (**a**): sifted flour in 9 in. (23 cm) pan; (**b**): collection of flour pellets; (**c**): separating pellets in a sieve stack; (**d**): weighing flour pellets.

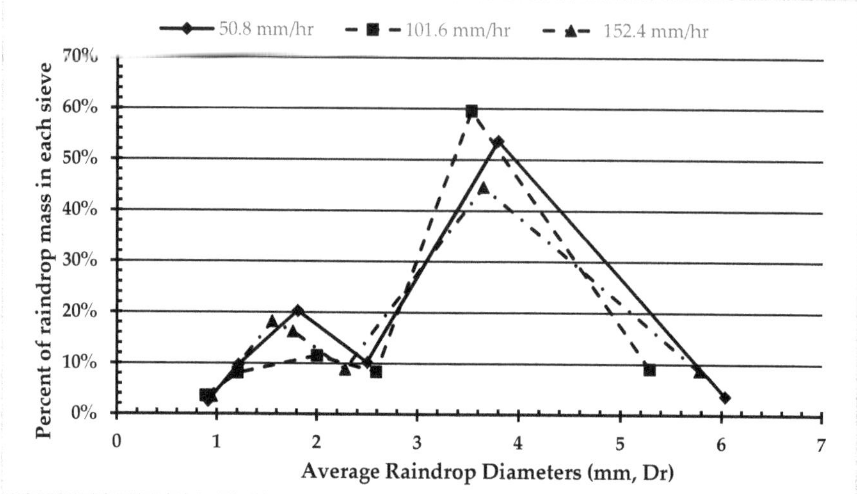

Figure 5. Percent of raindrop mass versus average raindrop diameter, D_r, in each sieve for three rainfall intensities.

Following the calculation of the raindrop size distribution, the kinetic energy generated at each rainfall intensity was calculated to determine rainfall energy, *E*. First, the raindrop fall height was determined by holding a surveyor's rod vertically in front of the center of a single sprinkler riser, extended above the height of the sprinkler nozzles, while the riser was operational. The wetted height was recorded as the average fall height for the raindrops. Next, using the average raindrop diameters computed from the flour pan method, the average volume of the raindrops was calculated using Equation (8):

$$V_{avg} = \frac{4\pi}{3}\left(\frac{D_r}{2}\right)^3 \tag{8}$$

where V_{avg} = average volume of raindrops (mm^3), and D_r = average diameter of raindrops (mm). The diameter of the drops is used with Figure 6 to determine the velocity of the drops falling from the height of the rainfall simulator and the terminal velocity. From this, the values for kinetic energy were calculated using Equation (9):

$$KE = 0.5mv^2 \tag{9}$$

where, KE = kinetic energy (J), m = average mass of raindrop (kg), and v = velocity of raindrop (m/s).

The final step of the calibration process was to calculate the erosion index using Equation (4).

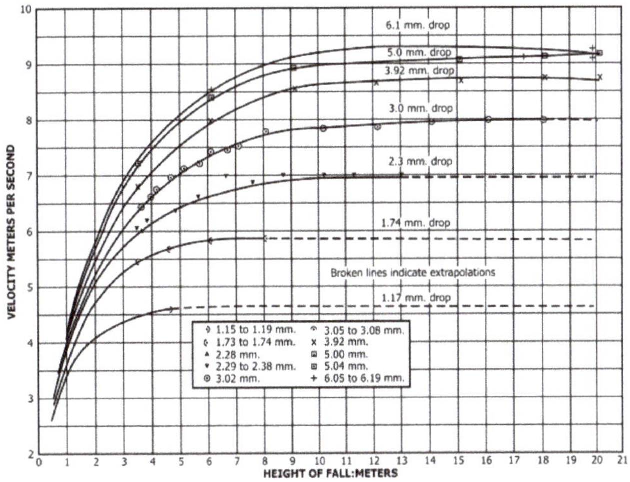

Figure 6. Fall velocity of raindrops as function of raindrop size and fall height [22]. Note: 1.0 ft = 0.305 m.

4. Results and Discussion

Calibration experiments were conducted to provide a means to quantify the performance of the rainfall simulator and determine if the apparatus is capable of simulating rainfall with characteristics similar to natural rainfall on a consistent basis. The methods and procedures previously discussed produced a multitude of data in the form of rainfall depth measured from each of the 29 rain gauges after each calibration test. The data from each test were analyzed to determine the average rainfall intensity and CUC. Finally, the values calculated from the calibration tests for each target rainfall intensity were averaged to provide a generalized report on the performance of the rainfall simulator in terms of experimental rainfall intensity and uniformity of rainfall distribution.

To validate the calibration process, a minimum of ten calibration tests for each intensity were conducted. If the standard deviation was less than or equal to 2.54 mm/hr (0.10 in./hr), testing efforts would proceed to the next interval. A maximum deviation of 2.54 mm/hr was set as the realistic limit for the simulator performing in a consistent and repeatable fashion. A total of 30, 15-min calibration tests were performed. The results from the calibration tests for all test intervals are summarized in Table 3. The flow rate column represents the sum of the flow rates provided by the nozzles at a constant pressure of 41.4 kPa (6.0 psi).

Table 3. Calibration summary for all test intervals.

Test Intervals [1]	Average Rainfall Intensity (mm/hr)	Sample Size	Standard Deviation (mm/hr)	Target Intensity (mm/hr)	Error (%)
1	52.83	10	1.02	50.8	4.00
2	104.65	10	1.52	101.6	3.00
3	154.18	10	1.78	152.4	1.17

Note: [1] test rainfall intervals are shown on Figure 6, 1.0 gal = 3.79 L, 1.0 in. = 25.4 mm.

After analyzing the values in Table 3, it was concluded that the rainfall simulator consistently produced rainfall intensities slightly higher than the theoretical target. According to Meyer [3] the experimental intensities should only vary from the theoretical intensities by a few percent. For the purpose of this study, the benchmark was set at 5.0%. Although the average rainfall intensities were higher than the theoretical target, the standard deviation between the 30 calibration tests was only 1.78 mm/hr (0.07 in./hr). This result ensured that the rainfall simulator was producing repeatable results in terms of rainfall intensity for all test intervals. Figure 7 graphically illustrates that the experimental rainfall intensities calculated during calibration were typically slightly higher than the theoretical targets. The intensity produced by the rainfall simulator follows a linear pattern based on the total flow rate in the sprinkler heads. The R^2 value quantifies how accurately the trend line fits the data. With a R^2 value of 0.994, the linear trend line serves as a reliable means for estimating flow rates and corresponding nozzle sizes required to simulate specific rainfall intensities.

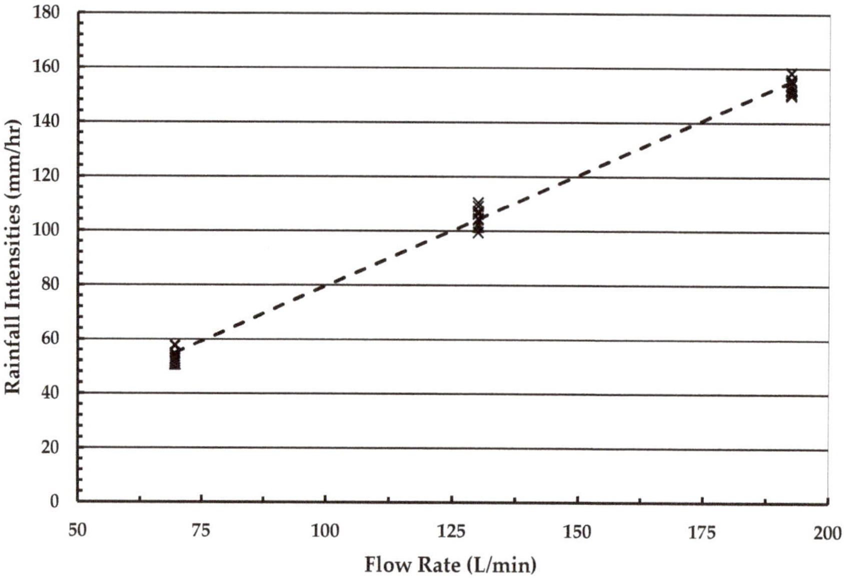

Figure 7. Experimental vs. theoretical rainfall intensities at different flow rates.

The average experimental rainfall intensities were used to calculate the *EI* using Equation (6) and to compare against theoretical values. *EI* is used in calculating the rainfall-runoff erosivity factor (R-factor) used in the RUSLE calculations for expected erosion over a given area. The R-factor is used to quantify the erosive energy of rainfall associated with specific storm events. The results from this analysis are presented in Table 4.

Table 4. Experimental vs. theoretical erosion index (EI) values.

Experimental Intensity (mm/hr)	Target Intensity (mm/hr)	Experimental Erosion Index	Target Erosion Index	Percent Error (%)
52.83	50.8	2169	2073	4.60
104.65	101.6	4510	4376	3.07
154.18	152.4	6669	6592	1.17

The calculated values in Table 4 correspond with the results from Figure 6. The higher rainfall intensities produced by the rainfall simulator result in greater erosive potential on the test slope. The ensuing result is that higher rates of soil erosion are generated by the simulated rainfall versus what should be expected from the actual storm event. However, as the rainfall intensities increase, the relative error between the experimental and theoretical values decrease.

To aid in the visualization of the uniformity of rainfall distribution for each test interval, raster surfaces showing rainfall intensity were generated using AutoCAD Civil 3D^{TM} and overlaid on an aerial photo of the test plot as shown in Figure 8. For each test interval, the rainfall intensities,

Figure 8a–c, were greatest in the middle of the test plot and lowest at the bottom of the test plot. The average uniformity of rainfall distribution for all tests performed ranged between 87.0 to 87.7%.

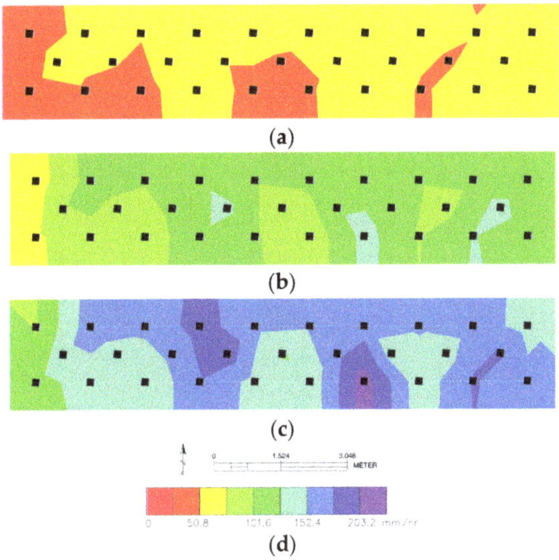

Figure 8. Rainfall intensity raster surfaces from calibration testing. (**a**) intensity–test interval 1; (**b**) intensity–test interval 2; (**c**) intensity–test interval 3; (**d**) legend.

The rain drop diameters produced by the simulator were calculated using the flour pan method. The average drop diameter was then used to calculate the average mass of the rain drops (Table 5). At each intensity, the calculated drop diameter was smaller than the theoretical value. Smaller diameter raindrops are produced when pressurized flow is discharged through small nozzle openings.

Table 5. Drop size distribution testing and kinetic energy of raindrops.

Rainfall Intensity, mm/hr (in./hr)	50.8 (2.0)	101.6 (4.0)	152.4 (6.0)
Average Drop Diameter, mm (in.)	2.39 (0.094)	2.58 (0.102)	2.35 (0.093)
Theoretical Drop Diameter, mm (in.)	2.53 (0.100)	2.87 (0.113)	3.09 (0.122)
Percent Error (%)	5.53	10.10	23.94
Average Drop Mass, mg (lbs)	7.13 (1.57×10^{-5})	8.97 (1.98×10^{-5})	6.77 (1.49×10^{-5})
Velocity of Drop, m/s (ft/s)	6.2 (20.3)	6.3 (20.7)	6.2 (20.3)
Kinetic Energy, J (lbs-ft2/s2)	7.24×10^{-5} (1.71×10^{-3})	8.37×10^{-5} (1.98×10^{-3})	7.00×10^{-4} (1.66×10^{-4})

Note: 1.0 in. = 25.4 mm.

The values for average drop mass calculated previously were used to determine the experimental kinetic energy generated by the rainfall simulator (Table 5). Values for rain drop velocity were estimated based on the diameter of the drop and the height from which the drops fell. In reality, the velocity of the drops is greater than estimated since the drops are projected from the sprinkler head with an initial outward and downward vector velocity. However, the actual velocities of raindrops were not quantified in this study.

As shown in Table 5 above, the measured average drop diameter (D_{50}, in mm) for each interval was less than the theoretical drop diameters. However, the experimental drop diameters are consistent with other pressurized rainfall simulators as detailed in Bubenzer [28]. As shown in Bubenzer [28], the median drop diameter for pressured simulators ranged from 0.6 to 2.6 mm (0.02 to 0.10 in.), with the majority of the simulators producing a median drop diameter of 2.1 mm (0.08 in.).

For each rainfall intensity, the kinetic energy of a single raindrop is negligible. However, when combined with the energy of the thousands of other raindrops impacting the slope each second, the summation of this energy would be much more considerable.

4.1. Bare Soil Control Testing

Once the rainfall simulator was calibrated, the next phase of the study involved performing a series of bare soil control tests for the design storm as specified in ASTM D6459-15. The tests consisted of a 60-min rainfall simulation with three separate intensities of 50.8, 101.6, and 152.4 mm/hr (2.0, 4.0, and 6.0 in/hr) for 20 min each. For these control tests, the soil tested was classified as a sandy loam as per the USDA soil texture triangle. The particle size distribution is shown below in Figure 9.

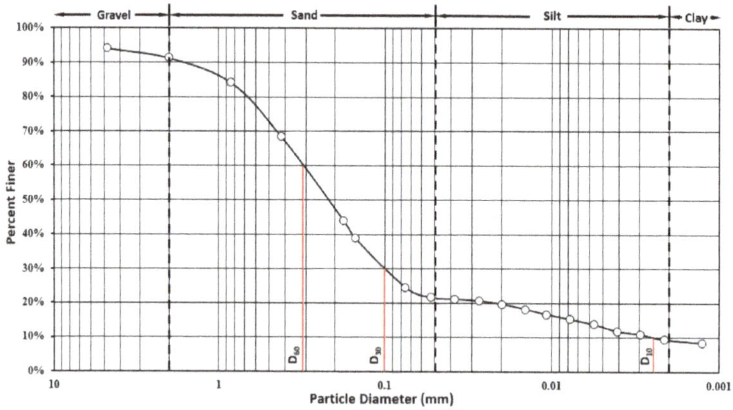

Figure 9. Particle size distribution for control soil.

Prior to performing each simulation, the test slope was prepared by tilling to a minimum depth of 10.2 cm (4.0 in.) and then compacting through the use of a lawn roller apparatus. The moisture content and compaction of the soil is then determined using the procedures outlined in ASTM D2937-10. During the simulation, grab samples of the runoff generated from the plot were captured at a maximum of every three minutes beginning once runoff began exiting the plot and continuing until the runoff ceased.

Summary plots are provided below in Figures 10 and 11 for both turbidity and total sediment concentration, respectively, for the four bare soil control tests conducted. A summary of the results is shown below in Table 6.

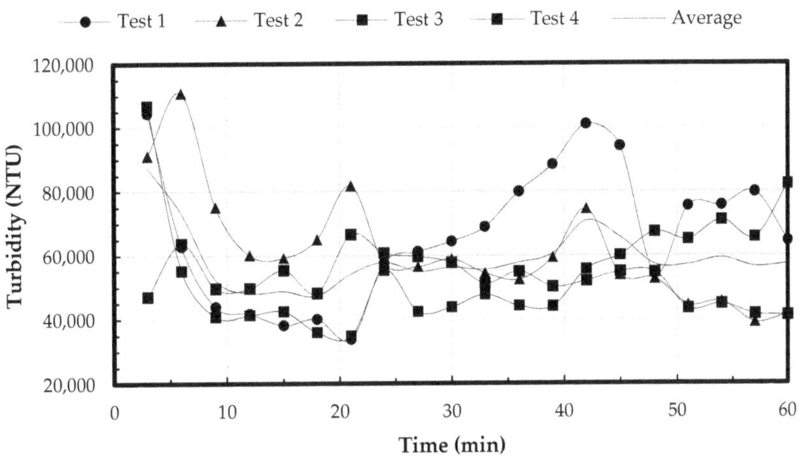

Figure 10. Turbidity versus time for control tests.

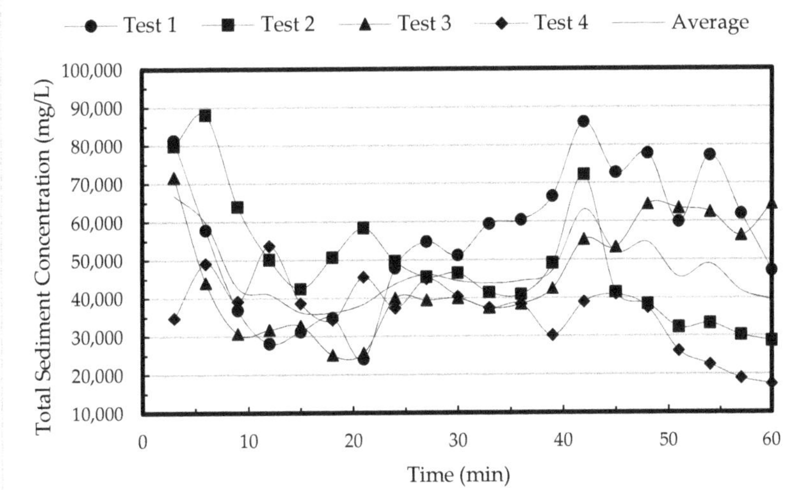

Figure 11. Total sediment concentration versus time for control tests.

Table 6. Summary of bare soil control test results.

Test Parameters	Test 1	Test 2	Test 3	Test 4	Average
Rainfall Depth (mm)	103.63	103.63	104.90	102.87	103.76
Compaction (%)	87.25	87.57	86.99	85.67	86.87
Moisture Content (%)	15.08	16.08	18.53	16.82	16.63
Catch Basin Sediment (kg)	329.16	322.87	341.34	346.83	335.05
Total Soil Loss per Unit Area (kg/ha)	110,819	108,699	113,644	116,770	112,483
Average Sediment Concentration (mg/L)	55,872	49,178	47,323	36,326	47,175
Average Turbidity (NTU)	66,493	61,683	56,119	52,374	59,167

4.2. Use of the Erosion Index Equation for Simulated Rainfall

It should be considered that the erosion index described in Equation (4) was developed from analyzing soil erosion resulting from years of naturally occurring rainfall events. This results in an assumption of naturally occurring drop size distribution and the raindrops falling at terminal velocity. For predicting naturally occurring erosion rates based upon this Equation, these assumptions are relatively valid for in-field conditions. However, for synthetic rainfall produced by rainfall simulators, these conditions are difficult to create. For instance, based upon Figure 6, a 3.0 mm (0.12 in.) raindrop has to fall approximately 16.0 m (52.5 ft) in a wind-free environment before reaching terminal velocity. ASTM D6459-15 only requires a minimum fall height of 4.26 m (14.0 ft). The nearest drop size plotted for this height in Figure 6 is 1.17 mm (0.05 in.) drop diameter, which is smaller than the average drop size produced by most rainfall simulators used for erosion testing. This issue is addressed in ASTM D 6459-15, but only minimally. After calculating the erosion index using Equation (4), the standard specifies that the results of this calculation must then be corrected for the kinetic energy of the drops that are falling at less than terminal velocity. Since no further guidance in the standard is provided for correcting the results of Equation (4), it is left to the user to determine how to adjust.

However, adjusting the output of Equation (4) as the ASTM standard stipulates may not be necessary and potentially improper. It can be seen from this previous discussion that intensity is the correlative variable that helps define the energy of a naturally occurring storm event using Equations (3), (4), and (9). However, should "correcting Equation (4) for kinetic energy" be performed since the simulators are not producing naturally occurring rainfall energy? The point of using Equations (3), (4), and (9) is to bypass determining the kinetic energy of each storm directly by using intensity as a means of estimating energy. To be able to adjust *EI* from Equation (4), the kinetic energy from the simulated storm must be known. Therefore, since the storm is not naturally occurring and may not be adequately represented by Equation (4), it may be more prudent to simply use the kinetic energy calculations from Equation (9) and directly calculate EI_{30} using the simulator's actual measured kinetic energy, instead of correcting Equation (4) that represents naturally occurring kinetic energy. These concepts require further evaluation and research, which is beyond the scope of this paper.

5. Summary and Conclusions

The purpose of this research study was to design and construct a large-scale rainfall simulator capable of repeatedly simulating rainfall with characteristics similar to natural rainfall. The design for the rainfall simulator was largely based on existing designs in ASTM D6459-15. However, changes were made due to the lack of specified products available for purchase and a desire to improve upon the existing ASTM standard testing method. Changes included: (1) using Nelson Irrigation PC-S3000 sprinkler heads in lieu of nozzles, no longer commercially available, stated in ASTM D6459-15) and (2) substituting solenoid valves in lieu of manual ball values. Discussion into the proper use of ASTM specified equations and their appropriateness were also introduced.

In accordance with ASTM D6459-15, rainfall intensities of 50.8, 101.6, 152.4 mm/hr (2.0, 4.0, and 6.0 in./hr) were simulated. Thirty, 15-min calibration experiments were conducted to determine the average experimental rainfall intensities and uniformities, drop size distribution, and erosive energy produced by the rainfall simulator. The experimental values were then compared with their corresponding theoretical targets to determine if the apparatus was adequately simulating natural rainfall. The theoretical target rainfall intensities for this study were 50.8, 101.6, and 152.4 mm/hr (2.0, 4.0, and 6.0 in./hr). The average experimental rainfall intensities produced by the rainfall simulator were found to be 53.8, 105.9, and 154.2 mm/hr (2.12, 4.17, and 6.07 in./hr), respectively. The uniformity of the rainfall distribution, quantified using CUC, was calculated to range between 87.0 to 87.7%. The corresponding average drop size were calculated to be 2.39, 2.58, and 2.35 mm (0.094, 0.101, and 0.093 in.), respectively. These results indicate that the rainfall simulator provides repeatable rainfall intensities, achieves uniformity, and produces consistent drop sizes.

As shown in Figures 9 and 10 above, the simulator was also used to perform a series of bare soil control tests to establish a baseline for future evaluation of hillslope erosion control products. The results summarized in Table 6 as well as the visual inspection of the test plot after each control test provide evidence that the rainfall simulator produced erosion results consistent within expected ranges. Visual inspections provided evidence of the occurrence of consistent splash erosion, sheet erosion, and rill erosion patterns; as well as consistent sediment yield, turbidity, and sediment concentrations measurements. Future research is also planned to test and evaluate various erosion control measures (i.e., rolled erosion control products, hydromulch, conventional mulching practices, etc.) on the test slope and to compare the performance and effectiveness of each respective practice in reducing erosion.

Author Contributions: M.D.R. conducted the simulations and analysis of the results, prepared the manuscript draft. M.A.H. designed the rainfall simulator and prepared the manuscript draft, B.F. conducted the simulations and analysis of the results, W.C.Z. supervised the design and simulations and revised the manuscript, X.F., W.N.D., and M.A.P. provided inputs on the writing, data analysis, and revised the manuscript. All authors made contributions to the study and writing the manuscript.

Funding: This research was funded by the Alabama Department of Transportation, Research Project 930-962.

Acknowledgments: The authors would like to thank the Alabama Department of Transportation for its sponsorship of this research. The authors gratefully acknowledge this financial support. The findings, opinions, and conclusions presented are those of the authors and do not necessarily reflect the view of the sponsors.

Conflicts of Interest: The authors declare no conflict of interest.

References

1. Novotny, V. *Water Quality: Diffuse Pollution and Watershed Management*; John Wiley & Sons, Inc.: New York, NY, USA, 2003.
2. United States Environmental Protection Agency (US EPA). *Stormwater Phase II Final Rule: An Overview*; US EPA: Washington, DC, USA, 2005.
3. Meyer, L.D. Rainfall Simulators for Soil Conservation Research. In *Soil Erosion Research Methods*; Lal, R., Ed.; Soil and Water Conservation Society: Ankeny, IA, USA, 1988; pp. 75–95.
4. Cullum, R.F. Electronic Controls for Rainfall Simulators. *Trans. ASAE* **1997**, *40*, 643–648. [CrossRef]
5. Hirschi, M.C.; Mitchell, J.K.; Feezor, D.R.; Lesikar, B.J. Microcomputer-Controlled Laboratory Rainfall Simulator. *Trans. ASAE* **1990**, *33*, 1950–1953. [CrossRef]
6. Mutchler, C.K.; Hermsmeier, L.F. A Review of Rainfall Simulators. *Trans. ASABE* **1965**, *8*, 67–68.
7. Paige, G.B.; Stone, J.J.; Smith, J.R.; Kennedy, J.R. The Walnut Gulch Rainfall Simulator: A Computer-Controlled Variable Intensity Rainfall Simulator. *Trans. ASAE* **2003**, *20*, 25–31.
8. Foltz, R.B.; Dooley, J.H. Comparison of Erosion Reduction Between Wood Strands and Agricultural Straw. *Trans. ASABE* **2003**, *46*, 1389–1396.
9. McLaughlin, R.A.; Brown, T.T. Evaluation of erosion control products with and without added polyacrylamide. *J. Am. Water Resour. Assoc.* **2006**, *42*, 675–684. [CrossRef]
10. Ming-Han, L.; Harlow, C.L.; Jett, M. Comparison of Field and Laboratory Experiment Test Results for Erosion Control Products. In Proceedings of the ASAE Annual International Meeting, Las Vegas, NV, USA, 27–30 July 2003.
11. Shoemaker, A.L. *Evaluation of Anionic Polyacrylamide as an Erosion Control Measure Using Intermediate-Scale Experimental Procedures*; Auburn University: Auburn, AL, USA, 2009.
12. Xiao, M.; Reddi, L.N.; Howard, J. Erosion Control on Roadside Embankment Using Compost. *Appl. Eng. Agric.* **2010**, *26*, 97–106. [CrossRef]
13. McLauglin, R.A.; Rajbhandari, N.; Hunt, W.F.; Line, D.E.; Sheffield, R.E.; White, N.M. The Sediment and Erosion Control Research and Education Facility at North Carolina State University. In Proceedings of the Soil Erosion Research for the 21st Century, Proceedings of the International Symposium, Honolulu, HI, USA, 3–5 January 2001.
14. Moore, I.D.; Hirschi, M.C.; Barfield, B.J. Kentucky Rainfall Simulator. *Trans. ASABE* **1983**, *26*, 1085–1089. [CrossRef]

15. Thomas, N.P.; El Swaify, S.A. Construction and Calibration of a Rainfall Simulator. *J. Agric. Eng. Res.* **1989**, *43*, 1–9. [CrossRef]
16. Sharpley, A.N.; Kleinman, P. Effect of Rainfall Simulator and Plot Scale on Overland Flow and Phosphorus Transport. *J. Environ. Qual.* **2003**, *32*, 2172–2179. [CrossRef] [PubMed]
17. Swanson, N.P. Rotating-Boom Rainfall Simulator. *Trans. ASABE* **1965**, *8*, 71–72. [CrossRef]
18. Meyer, L.D. Simulation of Rainfall for Soil Erosion Research. *Trans. ASABE* **1965**, *8*, 63–65. [CrossRef]
19. Shoemaker, A.L.; Zech, W.C.; Clement, T.P. Laboratory-Scale Evaluation of Anionic Polyacrylamide as an Erosion and Sediment Control Measure on Steep-Sloped Construction Sites. *Trans. ASABE* **2012**, *55*, 809–820. [CrossRef]
20. Kim, K.S.; Choi, J.D.; Kweon, K.S. Effect of PAM Application on Soil Erosion of a Sloping Field with a Chinese Cabbage Crop. In Proceedings of the ASAE Annual International Meeting, Sacramento, CA, USA, 29 July–1 August 2001.
21. Miller, W.P. A Solenoid-Operated, Variable Intensity Rainfall Simulator. *Soil Sci. Soc. Am. J.* **1987**, *51*, 832–834. [CrossRef]
22. ASTM. *Standard Test Method for Determination of Rolled Erosion Control Product Performance in Protecting Hillslopes from Rainfall-Induced Erosion*; ASTM: West Conshohocken, PA, USA, 2015.
23. USDA. *Predicting Soil Erosion by Water: A Guide to Conservation Planning With the Revised Universal Soil Loss Equation (RUSLE)*; USDA: Washington, DC, USA, 1997.
24. Bentley, W.A. Studies of Raindrops and Raindrop Phenomena. *Mon. Weather Rev.* **1904**, *32*, 450–456.
25. Laws, J.O.; Parsons, D.A. The Relation of Raindrop Size to Intensity. *Trans. Am. Geophys. Union* **1943**, *24*, 452–460. [CrossRef]
26. Nelson Irrigation. Available online: http://www.nelsonirrigation.com (accessed on 7 August 2019).
27. Sprague, C.J.; Sprague, J.E. Large-scale Performance Testing of Erosion Control Products. In Proceedings of the GeoAmericas 2012, Lima, Peru, 1–4 May 2012.
28. Bubenzer, G.D. Rainfall Characteristics Important for Simulation. In Proceedings of the Rainfall Simulator Workshop, Tuscon, AZ, USA, 7–9 March 1979; pp. 22–34.

water

MDPI

Article

Reconstruction of Seasonal Net Erosion in a Mediterranean Landscape (Alento River Basin, Southern Italy) over the Past Five Decades

Nazzareno Diodato [1] and Gianni Bellocchi [1,2,*]

1 Met European Research Observatory-International Affiliates Program of the University Corporation for Atmospheric Research, 82100 Benevento, Italy; scodalabdiodato@gmail.com

2 UCA, INRA, VetAgro Sup, Unité Mixte de Recherche sur Écosystème Prairial (UREP), 63000 Clermont-Ferrand, France

* Correspondence: gianni.bellocchi@inra.fr; Tel.: +33-4-4376-1601

Received: 28 August 2019; Accepted: 31 October 2019; Published: 4 November 2019

check for updates

Abstract: In the low Mediterranean basin, late spring and autumn rainfall events have the potential to increase discharge and transport substantial amounts of sediment soil (that is, the net soil erosion from a watershed). For the Alento River Basin (ARB), located in the low Tyrrhenian coast of Italy, we estimated changes of net erosion as dependent on the seasonality of antecedent soil moisture and its control on rainfall-runoff and erosivity. Based on rainfall and runoff erosivity sub-models, we developed a simplified model to evaluate basin-wide sediment yields on a monthly basis by upscaling point rainfall input. For the period 1951–2018, the reconstruction of a time series of monthly net erosion data indicated a decreasing trend of the sediment yield after 1991. Revegetation and land abandonment that occurred in the last decades can explain such a decrease of net erosion, which occurred even when rainfall erosivity increased. This response, obtained at the basic scale, does not exclude that rapidly developing mesoscale convective systems, typically responsible for the heaviest and most destructive rainfall events in the ARB, can affect small catchments, which are the most vulnerable systems to storm-driven flash floods and soil erosion hazards during soil tilling in spring and at beginning of autumn.

Keywords: erosive rainfall; parsimonious modeling; river basin; soil erosion

---

1. Introduction

Environmental changes are a prominent topic for Earth and environmental sciences, but its importance increases during crucial changes and different types of climate extremes that potentially lead to crises of some kind [1,2]. Extreme climate events are often associated with land degradation [3,4]. Soil erosion, in particular, is a pervasive form of soil degradation and a matter of increasing concern because of its implications for food security with the rapidly increasing world population [5]. Modeling processes that produce geomorphological hazards require understanding of how landscape components respond to forced conditions of land use change and to the climatic regime [6,7]. This is valuable to inform the assessment of future planning [8,9], but soil erosion monitoring systems tracking downstream sediment movement may be costly, and require focused efforts to manage land and water resources [10]. Because of this cost, modeling is playing an increasingly significant role [11]. This applies to the quantification of sediment dynamics, which is key to Earth-system science as documented in geology [12], biogeochemistry [13], and human activities [14]. It is also key to advancing our quantitative understanding and predictive capabilities of regional and sub-regional sediment fluxes. In the last decade, for instance, some geomorphological studies of long-term scale have affected the coastal areas of Southern Italy [15,16], which is the focus of this study, but research is still needed, given the differences in the responses at monthly and annual scales. Especially in mountainous agricultural areas, hydro-geomorphological degradation processes represent a complex issue, which manifests in a variety of phenomena [17]. Widespread modeling approaches are crucial in assessing climate variability and land cover [18,19], two important factors affecting the environmental sustainability of landscape systems [20–22]. The latter are dynamic and sensitive, and highly controlled by a set complex of climatic, geomorphic, and ecologic processes [23]. Sediment rates may be expected to change in response to changes in climate for a variety of reasons, responding both to the total amount of rainfall and to differences in rainfall intensity. However, the dominant variables appear to be rainfall intensity and energy, rather than rainfall amount alone. Nearing et al. [24] predicted that for every 1% increase in total rainfall, erosion rate would increase by only 0.85% if there was no corresponding increase in rainfall intensity. If both rainfall amount and intensity were to change together in a statistically representative manner, erosion rate would increase by 1.7% for every 1% increase in total rainfall, according to Pruski and Nearing [25]. Despite advances made in recent years, local-to-global modeling of sediment fluxes remains a research challenge [26]. For instance, numerical models have limitations for predicting basin sediment yield [27], especially over long timescales, and require access to powerful computer resources [28]. Their evaluation is also difficult because of the scarcity of measurements [29].

To deal with these issues, we propose an integrative methodology, based on the concepts of Foster et al. [30] and adapted from Thornes' [31] model (NETAM: Net Erosion Thornes-Adapted Model), offering a parsimonious interpretation of the relationship between hydrological data and basin-wide net erosion. We refer here to soil erosion by water, i.e., the result of rain detaching and transporting soil, either directly by means of rain splash or indirectly by rill and gully erosion. The capability to reproduce at basin scale the combined effects of hydro-climatological processes, including sediment transport, in the absence of distributed spatial and temporal data, relies on representation of the drainage basin as a homogeneous landform unit. In this way, the NETAM approach upscales point rainfall input data to area units where hydrological processes respond. This results in a long series of rainfall data from a single station (1951–2018 for the Alento River Basin (ARB)) that is a sufficient input for the parsimonious model. The use of a NETAM time-series model is thus motivated by its potential for capturing the significant and changing environment (including climate, vegetation cover, and erosive-resistance climate changes) with easily available data. Its evaluation in the ARB offers a unique opportunity to explore geomorphological processes in this Mediterranean fluvial basin.

2. Study Area

The Alento River Basin (ARB) is located between the Cilento, Vallo di Diano and Alburni National Park (Campania, southern Apennine), which is one of the largest Italian National Parks, stretching between 40°00′ and 40°30′ N, and 14°50′ and 15°00′ E (Figure 1a,b), with a total area of 428 km^2. The main weather station is located in Gioi Cilento (40°17 N, 15°13′ E), which holds the longest and most reliable hydrological data of the basin area (Figure 1c).

Altitudes range from sea level to Mt. Cervati (1898 m a.s.l.). Other peaks are located in the easternmost portion of the basin with Mount Sacro (1705 m a.s.l.), Mount Scuro (1610 m a.s.l.), and Mount Falascoso (1494 m a.s.l.). In southern Campania Region, three bioclimatic zones are present: The Mediterranean flat-hills along the coast, the pre-Apennines area, and inner hilly-mountainous zones. Precipitations vary from 600 to 1800 mm year^{-1}, depending on altitude and distance to sea (Figure 2a). The highest precipitation falls on Picentini Mounts, to the north, and Campano-Lucano Apennine, to the south (Figure 2a). With an enlarged view (Figure 2b), we can detect more resolute spatial variability also across the ARB. Here, precipitation varies from 800 mm year^{-1} on the valley to 1400 mm year^{-1} on the mountains of the National Park.

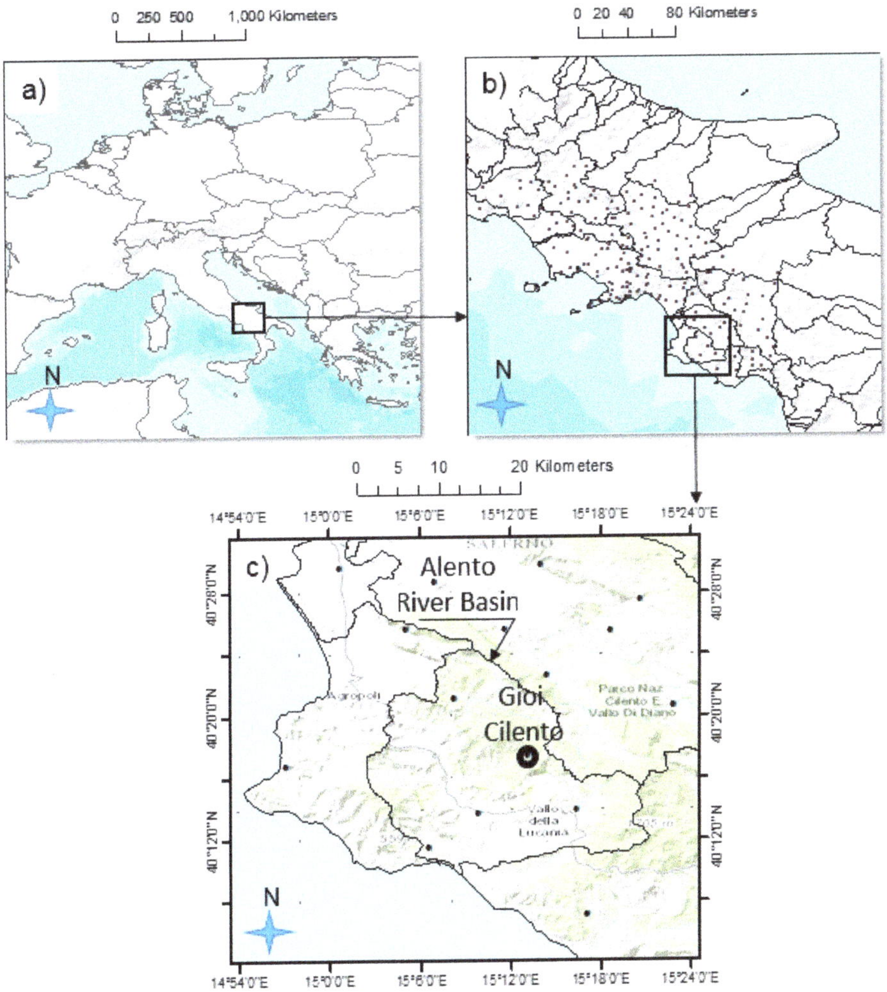

Figure 1. (**a**,**b**) Location maps of the Alento River Basin (ARB), and (**c**) hydrological data of the basin area. The station of Gioi Cilento (big dot) was used for the pluviometric elaboration and model-derived variables. All other stations (small dots) were used for mapping the annual mean precipitation and erosivity over the period 1951–2000.

River morphology is complex in the study region. The upstream part of the basin presents a narrow alluvial valley with steep slopes. Downstream, however, the river makes a turn towards the south. Subsequently, the river assumes a braided configuration down until the reservoir of Piano della Rocca, in the commune of Prignano Cilento (40°20′ N, 15°04′ E). Further downstream, the river mostly takes a meandering character. The geological nature of the rocks is dominated by the "Flysch of the Cilento" (i.e., limestone and silicoclastic substrata), wherein the main river basins (Alento, Calore, Mingardo, Bussento) are established [32]. Overall, the basin area is not prone to gully erosion, as it is dominated by erosion-resistant lithologies. With the only exception of the far northern/north-eastern part of the catchment, which is characterized by the presence of limestones pertaining to the Apennine Chain, these formations are quite homogeneous in hydrogeological terms and may be merged into a single hydrogeological complex of arenaceous–marly–clayey formation, which is relatively poorly permeable.

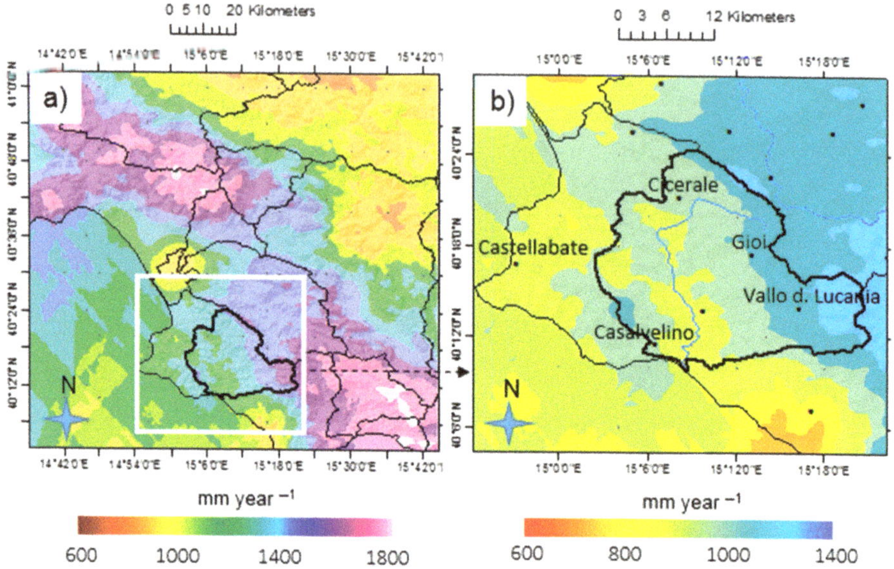

Figure 2. (**a**) Annual mean precipitation map over southern Campania Region and (**b**) relative zoom for the Alento River Basin upon the period 1951–2000. The maps were produced with ESRI-ArcGIS via Geostatistical Analyst (Lognormal Detrended Ordinary Cokriging with altitudinal covariate used for the purpose).

3. Materials and Methods

3.1. Data Collection

Daily rainfall data for the period 1951–2000 were collected from the rain gauge network of the *Servizio Idrografico and Mareografico Nazionale* (SIMN, National Hydrographic and Marine Service) [33], nowadays *Rete Mareografica Nazionale* (National Tidegauge Network, http://www.mareografico.it), continued by *Centro Funzionale Multirischi Protezione Civile della Regione Campania* (Multirisk Functional Centre of Civil Protection–Campania Region, http://centrofunzionale.regione.campania.it/#/pages/dashboard). However, for sediment data, only a long-term (1951–2000) average value was available [34]. Under this limited calibration condition, credibility of final output estimates was founded on the sub-model validation. The model was thus calibrated against long-term average net erosion and then validated for its erosivity and runoff sub-models, using monthly aggregated data, as determined in the ARB from the sub-periods 2002–2008 (RUSLE based erosivity at Gioi Cilento [35]) and 1958–1973 (SIMN measurements at the outlet of the ARB), respectively.

Monthly vegetation cover fraction was assessed with Normalized Difference Vegetation Index (NDVI) data, as derived from the GIMMS–KNMI Climate Explorer platform (http://climexp.knmi.nl), and rearranged to characterize the inter-annual evolution [16]. Olive orchards and sclerophyllous Mediterranean vegetation prevail along the coast, whereas forest landscape is dominant in the inner area, mainly represented by *Quercus cerris* or *Fagus sylvatica* woods. Not negligible is also the presence of riparian forest cover, dominated by *Salix alba*, *Populus nigra*, *Populus alba*, and *Alnus glutinosa*. Smallholder agriculture (arable land and orchards) dominates, sustained by mechanization, road infrastructure, availability of groundwater stocks, and water storage for irrigation purposes.

3.2. Net Erosion Model

Net soil water erosion is a measure of average sediment yield (soil net erosion) occurring basin-wide over time (Figure 3a), resulting from the sum of the sediment produced by all erosional sources, including overland flow, ephemeral gully, and stream channel areas [28], minus the amount of sediment deposited on such transfer zones and on the valley floodplains. The result is the amount of sediment conveyed downstream to the outlet of the basin. Four environmental factors determine the amount of water erosion and sedimentation. They are climate, soil, topography, and land-use, which operate independently and interactively. Basic characteristics and spatio-temporal features are thus taken into account in a hierarchical structure for discovering erosional phenomenon. In particular, the evolution over time of net erosion reflects the magnitude and frequency of individual storm events, which are nested within larger events occurring on different time scales [21].

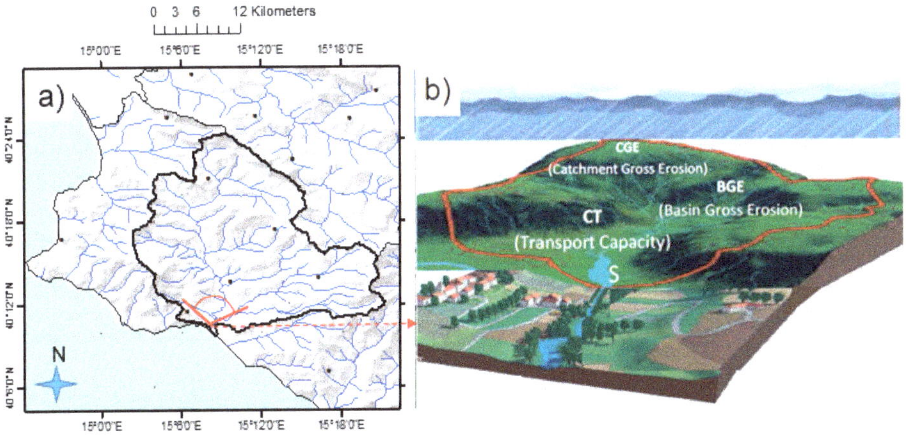

Figure 3. (**a**) Orography of the ARB with the fluvial drainage network facing the Tyrrhenian Sea and (**b**) three-dimensional (3-D) exemplary view of landscape hydro-geomorphological processes in a nested scheme for the ARB.

Figure 3b outlines the role played in sediment transport by mesoscale rainstorms accounted at the basin scale (BGE, basin gross erosion), while also assuming that the distribution of local showers play an important role in determining torrential flows rich in sediment in the individual river catchments of the basin (CGE, catchment gross erosion).

The model structure suggests that spring–summer (May to September) precipitation is an important factor to estimate the relative contribution of individual catchments (upper tributaries river) to the sediment (CGE) moving within the basin drainage system. In contrast, winter precipitation mostly contributes to basin-wide transient response (BGE within lower tributary river).

Since the procedure for determining rainfall erosivity suggested by Wischmeier and Smith [36] is applicable to the computation of annual erosion, its use to estimate soil loss from single storms would imply considerable errors [37] and motivate a reinterpretation of the original formulation. Foster et al. [30] and Thornes [31] elaborated the concept of the balance between driving and resisting forces in sediment budget. We further arranged this solution to model net erosion on a monthly basis (NETAM, Mg km^{-2} month^{-1}) as:

$$NETAM = A \cdot \left[k \cdot S^n \left(\alpha \cdot R_S^u + \beta \cdot R_Q^m \right) \cdot e^{-v \cdot VCF} \right] \quad (1)$$

where the term within round brackets is the modified Foster algorithm; R_S is the rainfall-erosivity indicator associated with splash erosion; R_Q (mm) is the runoff term, associated with transport erosion;

S (m^{-1}) is the mean slope of the basin; the erodibility coefficient (lithology factor) $k = 0.0145$ and the shape parameters (which play an adjusting role on the model inputs) $n = 2$ and $m = 2$ were arranged from Wainwright and Mulligan [11], and u was determined by calibration; exp(-$v \cdot VCF$), with $v = 0.07$, is the exponential vegetation function [31], with VCF (%) being the vegetation cover fraction [16]; A and α are erosivity scale coefficients, whose values were determined by calibration.

Our approximation is that hydraulically rough and vegetated surfaces reduce flow velocity and, hence, soil interril transport capacity [28]. This is reflected in the low values attributed to parameters k and v in Equation (1). Then, as canopy cover reduces soil detachment caused by raindrop impact, it also reduces interrill sediment transport capacity by attenuating raindrop impact. Based on this understanding, the power of rainfall as prevailing storm erosivity in summer and autumn is captured by the daily rainfall term of R_S, while in winter and spring, runoff is captured by the monthly rainfall terms of R_Q. In the ARB, predominant water erosion derives from interactions between the detachment on hillslope areas caused by water drops falling on soil, and successive runoff towards downslope up to flow in the drainage networks. This linkage of processes occurs within a fluctuating and continuous interplay of disturbing and resistance forces. In this way, soil erosion by water mostly occurs when the detachment of particles and their subsequent transportation experience a greater driving force than the force binding particles into the vegetated slope. With all these processes, rainfall is used by nature as both a driving and a resisting factor. To better detail this, firstly the erosive influence of rainfall increases with water amount, intensity, and runoff; secondly, and opposing this influence, the protective effect of vegetation increases with precipitation amount.

To further explain the single terms of Equation (1), arranging from Diodato and Aronica [38], we obtain:

$$R_S = \sqrt{dx} \cdot (dx \cdot f(jm)) \tag{2}$$

where dx is the daily maximum rainfall (mm) in each j month; the scale-factor $f(j_m)$ is as follows:

$$f(jm) = \left(1 - 0.45 \cdot \cos\left(6.28 \frac{j - 2.5}{22 - j}\right)\right) \tag{3}$$

The semi-parametric function $f(j_m)$ modulates the intra-seasonal storm intensity proxy during rainfalls.

The following R_Q term represents, instead, the erosivity mostly associated with runoff erosion:

$$R_Q = \left(p + p_{j-1}\right) \cdot w \tag{4}$$

where p is the amount of rainfall (mm) in the current month and p_{j-1} (mm) is the rainfall in the previous month; w is an indicator of soil humidity, in the form of a semi-parametric function, to modulate the intra-seasonal humidity after precipitation:

$$w = \left(0.5 + 0.4 \cdot \cos\left(6.28 \frac{j + 0.5}{24 - j}\right)\right) \tag{5}$$

4. Results and Discussion

Overall results of the calibrated model (1951–1990) and sub-model (rainfall erosivity and runoff) validation are first presented, followed by the model-based reconstruction of net erosion data for the period 1951–2018. The long-term trend is discussed at the annual scale, before highlighting the net erosion variability at the monthly scale.

4.1. Model Calibration

For the calibration period 1951–1990, over which long-term annual mean of net erosion data was available for the ARB, the values of the coefficients $u = 2$, $A = 1500$ (which converts values of eroded soil from mm to Mg km^{-2}), and $\alpha = 0.1$ in Equation (1) were obtained by approximating the model output to the silting value determined experimentally from the degree of filling of the dam of Piano della Rocca (490 Mg km^{-2} year^{-1}), covering 24% of the entire basin [34]. If the figure of 490 Mg km^{-2} year^{-1} for the period 1951–1990 is extrapolated for the whole of the basin, then the overall erosion rate calculates to 2042 Mg km^{-2} year^{-1}. The calibrated estimate was 2041 Mg km^{-2} year^{-1} for the same period.

4.2. Semi-Quantitative Validation

To ensure that the model serves its intended purpose, a semi-quantitative verification with inter-monthly variability was done, since sub-models of Equations (2) and (4) do not include any scale parameter. Figure 4 shows the performance of these sub-models. In particular, Figure 4a displays that the rainfall-erosivity component is in agreement with RUSLE-based erosivity data [35]. Figure 4b also reflects a satisfactory performance between predicted and actual runoff data [33].

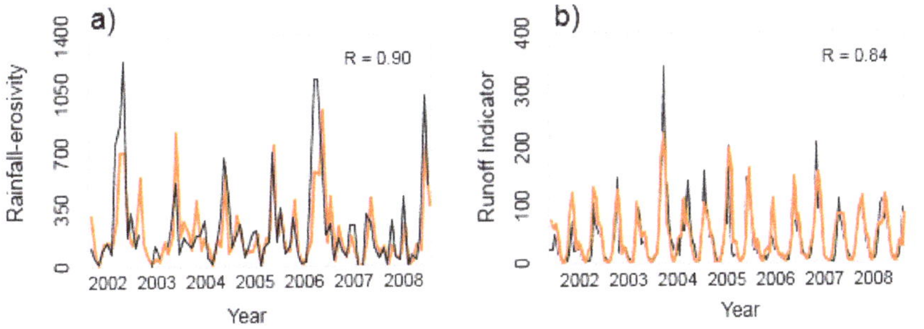

Figure 4. (**a**) Predicted erosivity indicator (orange curve, Equation (2)) and actual rainfall-erosivity (black curve, MJ mm ha^{-1} h^{-1} month^{-1}) for Gioi Cilento station (2002–2008), and (**b**) predicted runoff indicator (orange curve, Equation (4)) and actual runoff in mm (black curve) for the ARB (1958–1972).

This indicates that, at basin scale, net erosion is not the result of the runoff amount only, but of the combination of rainfall erosivity by both raindrop impact and surface runoff. As well, vegetation covers the soil during several months, over which erosion patterns may change [39].

In Figure 5, the model appears to correctly compute the main effects and trends associated with sediment yield, represented in this case by the sand extracted every year (proxy of the net erosion) in the Alento valley [40]. We evaluated the relative performance of the NETAM, without comparing the absolute estimates. Coevolution between material extracted at the valley of ARB (histogram) and simulated net erosion (blue curve) illustrates a substantial agreement, with the only exception of around 1985 (corresponding to the beginning of the construction of the dam in 1984 [34]).

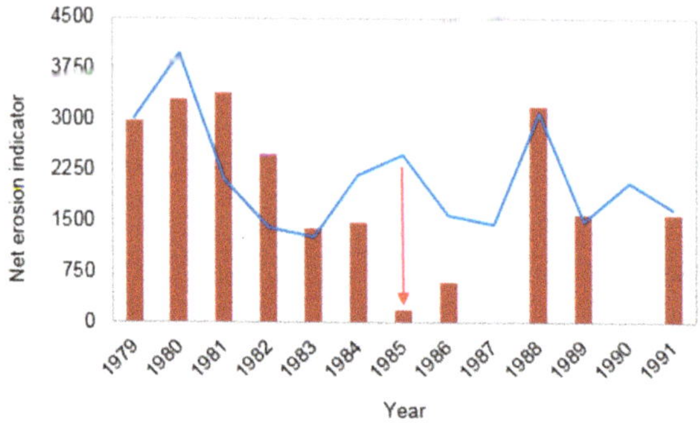

Figure 5. Coevolution time-series of extracted sediment in the Alento valley (histogram) and annual net erosion (blue curve, Equation (1), Mg km^{-2} year^{-1}) for the period 1979–1991. Red arrow indicates the disagreement occurring at the start of the dam construction.

4.3. Annual Net Erosion Reconstruction

Figure 6 shows the temporal evolution of annual sediment exports from ARB during the period 1951–2018, as calculated with Equation (1). Part of the estimated sediment was trapped by the dam built in 1994. However, no refinement was brought to the original dataset, as the sediment trapped in the reservoir is still the effective erosion that occurred from the several catchments composing the basin.

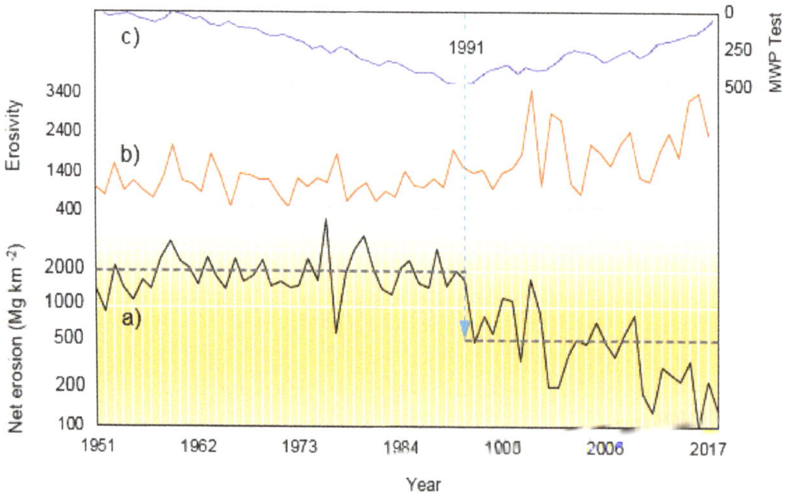

Figure 6. (**a**) Temporal evolution of the modeled—Equation (1)—annual amount of net erosion over the period 1951–2018 in the ARB (black curve), with the respective long-term mean values (bold dashed grey lines) before and after the change-point of 1991 (vertical dashed blue line). (**b**) Rainfall-erosivity in MJ mm ha^{-1} h^{-1} year^{-1} (orange curve) and (**c**) Mann–Whitney–Pettitt (MWP) test (blue curve) for the change-point detection.

Figure 6a, in particular, shows the actual evolution of net erosion that, after the first years with low erosional rate, reveals an increase from 1960 at a roughly constant trend that extended until 1990, before the change-point detected in 1991 (Figure 6c) with the Mann–Whitney–Pettitt test [41]. After this year, the sediment rate underwent a continuous irregular decrease until the end of the time-series.

Over the first period, 1951–1990, the average estimated net erosion value is of 2041 Mg km^{-2} year^{-1} (± 889 Mg km^{-2} year^{-1} standard deviation), while in the last period, 1991–2018, the average value stands at 568 (±436 Mg km^{-2} year^{-1} standard deviation), with a marked decrease of 1473 Mg km^{-2} year^{-1} compared to the previous period. This decrease is also accompanied by an amplification of the interannual variation coefficient of net erosion, which passes from 0.42 for the period before the change-point, to 0.58 for the following period. The growing seasonal irregularity of the precipitation is probably the major driver of the increasing interannual variability of soil erosion. Over 1991–1998, forest cover doubled and cropland roughly halved due to decades of land abandonment and reduction of human pressure [42], and this is likely the cause of decrease in net erosion during the period 1991–2018, although rainfall-erosivity kept on rising (Figure 6b). Thus, vegetation cover exerted a great resistance to the hydrological hazard, since vegetation underwent a general increase after 1990 [16]. However, during the most extremes hydrological events (e.g., precipitation at hours or sub-hourly scales), soil erosion in small catchments could represent a large risk for soil mobilization and transport, which can contribute to nutrient and organic carbon losses.

4.4. Net Erosion Monthly Variability and Timing

The use of monthly data helps summing up consecutive hydro-geomorphological events over an appropriate time scale, with respect to hydrological timing and crop growing cycles or scheduling of tillage practices. The modeled results, obtained over 68 years, show that there is a significant variation of sediment transport at the intra-seasonal scale in both past (1951–1990) (Figure 7a) and recent (1991–2018) periods (Figure 7b, grey bars). Almost half of suspended solid transport occurs in autumn (43% and 54%, respectively, in the two periods) and approximately one third of the annual flux occurs in winter (39% and 29%, respectively). However, the 95th percentile (Figure 7, empty bars) is distributed differently than to mean values, with more divergence in April and September (Figure 7, red bars). This divergence represents a high risk of soil erosion in correspondence to the months with tilled soil in both the periods, although results evidence a decrease of net erosion in all months during the recent phase 1991–2018 (Figure 7b).

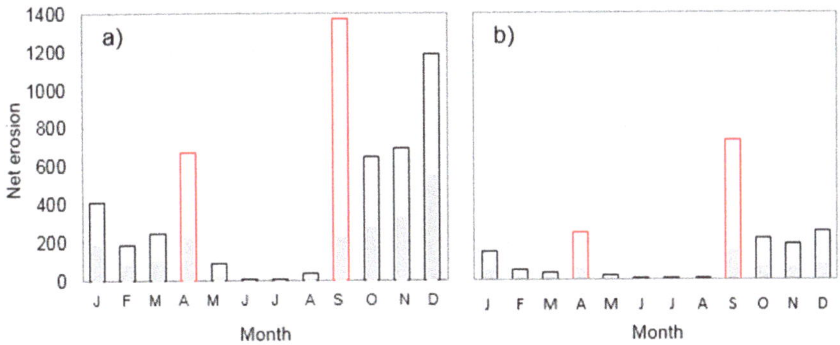

Figure 7. Seasonal evolution of modeled—Equation (1)—net erosion with mean values (grey bars) and 95th percentile (empty bars) during (**a**) 1951–1991 and (**b**) 1991–2018 for the ARB (in red bars, the erosional soil degradation hazard in April and September months, when storms occur with tilled soil).

The autumn season seems to maintain the primacy of erosion rates, in past as in recent times. Rizzi [43] documented disasters in the Alento coast in autumn and winter during past times.

In winter, rainfall and average sediment are significant, but most of the erodible particles are transported by the first floods of the preceding autumn. Spatial timeline of storminess also shows a decadal trend (Figure 8). The increasing trend has affected practically the entire basin, especially for storms of 24-h duration (Figure 8b). Then, the increased variability and amount of storms found at the Gioi Cilento station can affect the areas around the station. In particular, it is understood that an average increase of 10 mm per half a century affected the storms of 1-h duration (Figure 8a), and 10–20 mm the storms of 24-h duration (Figure 8b). The areas of the basin more interested from storm increases are those included along the transect zone around the villages of Cicerale (40°21 N, 15°08′ E), Gioi Cilento (40°17 N, 15°13′ E), and Vallo della Lucania (40°14 N, 15°16′ E).

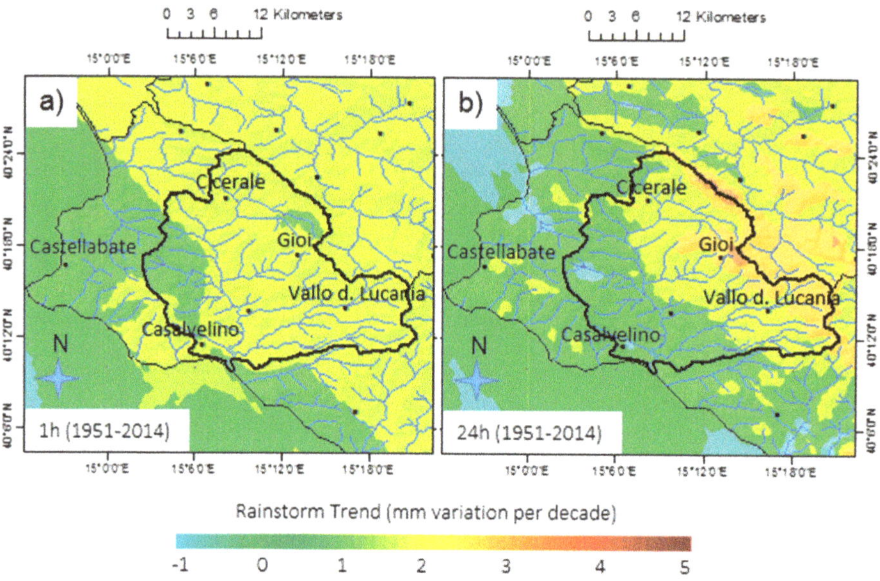

Figure 8. (**a**) 1-h duration rainstorm trend (mm decade^{-1}) and (**b**) as in (**a**) for 24-h rainstorm durations during the period 1951–2014 across the ARB (arranged from ESRI-Geostatistical Analysis via de-trended ordinary cokriging using as covariate the elevation).

5. Conclusions

Land use change has been recognized throughout the world as an important driver of climate-driven geomorphological processes, which may also trigger changes in carbon cycling [44]. Soil erosion rates may be expected to change in response to changes in climate and vegetation for a variety of reasons, the most direct of which is the change in the erosive power of rainfall and resistance forces, respectively. However, modeling rainfall-driven soil erosion rates is difficult because of the lack of long-term data in river basins. In particular, complex models are often not adequate to reconstruct net erosion (or sediment yield) changes because they require a considerable amount of high-resolution input data, not always available on long timescales. Thus, the use of parsimonious models offers an interesting possibility to reconstruct net erosion series on a monthly basis. This is what we have done with the NETAM, developed on the original Foster and Thornes algorithms, in a test site, the Alento River Basin (~400 km^2 in Southern Italy). Though the model developed for the ARB is not easily transferable for applications in other basins, it provided a peculiar and unique opportunity for modeling erosion responses to climate and land cover changes, where documented hydrological processes at basin scale also support input-data generation and interpretation of results. The ARB is a catchment with extensive natural areas. The development of agricultural and natural areas is favored by the presence of farming practices and a markedly seasonal climate. Thanks to the continuous observation of selected physical environmental variables, we were able to establish seasonal patterns of weathering processes and identify the factors that control rainfall erosivity and runoff and, in turn, net erosion. Cold and wet cycles in winter and wet and dry cycles in spring–autumn are the main processes involved in landscape weathering, thereby controlling slope development together with rainfall-related erosion processes. The main observed feature is the reaction of the ARB to all rainfall events. Hydrological events show high fluctuations of the suspended sediment by month-to-month, and by year-to-year, deriving from a heterogeneous temporal distribution related to seasonal variations of the hydro-climatic forcing (that is, surface erosivity and runoff) and the vegetation cover. In this way, NETAM values were obtained for the period 1951–2018 by using parsimonious erosion sub-models and land cover statistics from documented agrarian sources. We conclude that if pulses of sediment fluctuation in the ARB have always been driven mainly by natural climatic oscillations, then land abandonment and revegetation are the causes of the observed reduction of net soil erosion in the last decades. This study adds to a growing body of literature on the development of methodological frameworks and tools that could be used to outline scenarios of soil erosion and instability risks resulting from climate changes (e.g., increasing heavy rainfall events), and changes in land use and management practices in central Italy [45–47].

Author Contributions: N.D. conceived the study, performed the analysis, and drafted the manuscript. G.B. contributed to the analysis and wrote the final manuscript.

Funding: This is an investigator-driven study without any grant support.

Acknowledgments: We acknowledge the support of Mauro Biafore, Director of Centro Funzionale Multirischi Protezione Civile della Regione Campania (Naples, Italy), and Matteo Gentilella of the same center, who provided the rainfall data used in this study.

Conflicts of Interest: The authors declare no conflicts of interest.

References

1. Jentsch, A.; Beierkuhnlein, C. Research frontiers in climate change: Effects of extreme meteorological events on ecosystems. *Comptes Rendus Geosci.* **2008**, *340*, 621–628. [CrossRef]
2. Li, Z.; Fang, H. Impacts of climate change on water erosion: A review. *Earth-Sci. Rev.* **2016**, *163*, 94–117. [CrossRef]
3. Riebau, A.R.; Fox, D.G. Damage Assessment of Agrometeorological Relevance from Natural Disasters: Economic and Social Consequences. In *Natural Disasters and Extreme Events in Agriculture*; Springer Science and Business Media LLC: Berlin/Heidelberg, Germany, 2005; pp. 119–135.
4. Yin, J.; Gentine, P.; Zhou, S.; Sullivan, S.C.; Wang, R.; Zhang, Y.; Guo, S. Large increase in global storm runoff extremes driven by climate and anthropogenic changes. *Nat. Commun.* **2018**, *9*, 4389. [CrossRef] [PubMed]
5. Blaschke, P.M.; Trustrum, N.A.; Hicks, D.L. Impacts of mass movement erosion on land productivity: A review. *Prog. Phys. Geogr. Earth Environ.* **2000**, *24*, 21–52. [CrossRef]
6. Leeder, M.R.; Harris, T.; Kirkby, M.J. Sediment supply and climate change: Implications for basin stratigraphy. *Basin Res.* **1998**, *10*, 7–18. [CrossRef]
7. Coulthard, T.J.; Kirkby, M.J.; Macklin, M.G. Modelling geomorphic response to environmental change in an upland catchment. *Hydrol. Process.* **2000**, *14*, 2031–2045. [CrossRef]
8. Higgitt, D.L. A brief time of history. In *Geomorphological Processes and Landscape Change: Britain in the Last 1000 Years*; Higgitt, D.L., Lee, E.M., Eds.; Blackwell Publishers Ltd.: Oxford, UK, 2001; pp. 1–26.
9. Verstraeten, G.; Poesen, J. Factors controlling sediment yield from small intensively cultivated catchment in a temperate humidclimate. *Geomorphology* **2001**, *40*, 123–144. [CrossRef]
10. Trimble, S.W.; Crosson, P.U.S. Soil erosion rates—Myth and reality. *Science* **2000**, *289*, 248–250. [CrossRef]
11. Wainwright, J.; Mulligan, M. *Environmental Modelling: Finding Simplicity in Complexity*; John Wiley and Sons, Ltd.: Chichester, UK, 2004; p. 408.
12. Pelletier, J.D. A spatially distributed model for the long-term suspended sediment discharge and delivery ratio of drainage basins. *J. Geophys. Res. Space Phys.* **2012**, *117*, 02028. [CrossRef]
13. Syvitski, J.P.M.; Milliman, J.D. Geology, Geography, and Humans Battle for Dominance over the Delivery of Fluvial Sediment to the Coastal Ocean. *J. Geol.* **2007**, *115*, 1–19. [CrossRef]
14. Kettner, A.J.; Restrepo, J.D.; Syvitski, J.P.M. A Spatial Simulation Experiment to Replicate Fluvial Sediment Fluxes within the Magdalena River Basin, Colombia. *J. Geol.* **2010**, *118*, 363–379. [CrossRef]
15. Budetta, P. Landslide hazard assessment of the Cilento rocky coasts (Southern Italy). *Int. J. Geol.* **2013**, *7*, 1–8.

16. Romano, N.; Nasta, P.; Bogena, H.; De Vita, P.; Stellato, L.; Vereecken, H. Monitoring Hydrological Processes for Land and Water Resources Management in a Mediterranean Ecosystem: The Alento River Catchment Observatory. *Vadose Zone J.* **2018**, *17*, 1–12. [CrossRef]
17. Shrestha, D.; Zinck, J.; Van Ranst, E. Modelling land degradation in the Nepalese Himalaya. *Catena* **2004**, *57*, 135–156. [CrossRef]
18. Renschler, C.S.; Harbor, J.; Harbor, J. Soil erosion assessment tools from point to regional scales—the role of geomorphologists in land management research and implementation. *Geomorphology* **2002**, *47*, 189–209. [CrossRef]
19. Ramos, M.C.; Mulligan, M. Impacts of climate variability and extreme events on soil hydrological processes. In Proceedings of the EGS-AGU-EUG Joint Assembly, Nice, France, 6–11 April 2003. Abstract id. 11592.
20. Alexandrov, V.A. Vulnerability of agronomic systems in Bulgaria. *Clim. Chang.* **1997**, *36*, 135–149. [CrossRef]
21. Thomas, M.F. Landscape sensitivity in time and space—An introduction. *Catena* **2001**, *42*, 83–98. [CrossRef]
22. Reinhard, M.; Alexakis, E.; Rebetez, M.; Schlaepfer, R. Climate-soil-vegetation interactions: A case study from the forest fire phenomenon in Southern. In Proceedings of the EGS-AGU-EUG Joint Assembly, Nice, France, 6–11 April 2003. Abstract id. 2470.
23. Krishnaswamy, J.; LaVine, M.; Richter, D.D.; Korfmacher, K. Dynamic modeling of long-term sedimentation in the Yadkin River basin. *Adv. Water Resour.* **2000**, *23*, 881–892. [CrossRef]
24. Nearing, M.A.; Jetten, V.; Baffaut, C.; Cerdan, O.; Couturier, A.; Hernández, M.; Le Bissonnais, Y.; Nichols, M.H.; Nunes, J.P.; Renschler, C.S.; et al. Modeling Response of Soil Erosion and Runoff to Changes in Precipitation and Cover. *Impacts Glob. Clim. Chang.* **2005**, *61*, 1–11.
25. Pruski, F.F.; Nearing, M.A. Climate-induced changes in erosion during the 21st century for eight U.S. locations. *Water Resour. Res.* **2002**, *38*, 34-1–34-11. [CrossRef]
26. Cohen, S.; Kettner, A.J.; Syvitski, J.P. Global suspended sediment and water discharge dynamics between 1960 and 2010: Continental trends and intra-basin sensitivity. *Glob. Planet. Chang.* **2014**, *115*, 44–58. [CrossRef]
27. De Vente, J.; Poesen, J.; Verstraeten, G. The application of semi-quantitative methods and reservoir sedimentation rates for the prediction of basin sediment yield in Spain. *J. Hydrol.* **2005**, *305*, 63–86. [CrossRef]
28. Toy, T.J.; Foster, G.R.; Renard, K.G. *Soil Erosion: Prediction, Measurement, and Control*; John Wiley & Sons, Inc.: New York, NY, USA, 2002; p. 338.
29. Tarboton, D.; Pack, R.T.; Luce, C.H.; Istanbulluoglu, E. Modeling of the interactions between forest vegetation, disturbances, and sediment yields. *J. Geophys. Res. Space Phys.* **2004**, *109*, 01009.
30. Foster, G.R.; Meyer, L.D.; Onstad, C.A. A Runoff Erosivity Factor and Variable Slope Length Exponents for Soil Loss Estimates. *Trans. ASAE* **1977**, *20*, 683–687. [CrossRef]
31. Thornes, J.B. The interaction of erosional and vegetational dynamics in land degradation: Spatial outcomes. In *Vegetation and Erosion: Processes and Environments*; Thornes, J.B., Ed.; John Wiley & Sons: Chichester, UK, 1990; pp. 45–55.
32. Lapenna, M.R.; Rosati, L.; Salerno, G.; Villani, M.; Fascetti, S.; Filesi, L. Landscape planning and biodiversity conservation of river habitats require vegetation analysis and mapping: The case of Cilento National Park (Italy). In *Latest Trends in Energy, Proceedings of the 7th International Conference on Environmental and Geological Science and Engineering (EG '14); Proceedings of the 7th International Conference on Urban Planning and Transportation (UPT '14); Proceedings of the 3rd International Conference on Energy Systems, Environment, Entrepreneurship and Innovation (ICESEEI '14) Environment and Development, Salerno, Italy, 3–5 June 2014*; Griselda, J., Breton, C., Quartieri, J., Guida, M., Guida, D., Guarnaccia, C., Eds.; WSEAS: Sofia, Bulgaria, 2014; pp. 182–190.
33. SIMN. *Hydrological Annals*; Servizio Idrografico e Mareografico Nazionale: Rome, Italy, 1957–1972. (In Italian)
34. De Vita, P.; La Barbera, G. Studi per il progetto di gestione dell'invaso di Piano della Rocca sul fiume Alento (Campania Meridionale): Qualità delle acque e stima dell'interrimento. In Proceedings of the Atti del XXX Convegno di Idraulica e Costruzioni Idrauliche IDRA, Rome, Italy, 10–15 September 2006; pp. 1–16. (In Italian).
35. Chirico, G.B.; De Falco, M.; Diodato, N.; Romano, N.; Santini, A. Estimating rainfall erosivity from daily precipitation records in Campania region (Southern Italy). In Proceedings of the EGU General Assembly 2010, Vienna, Austria, 2–7 May 2010; p. 7084.

36. Wischmeier, W.H.; Smith, D.D. *Predicting Rainfall Erosion Losses from Cropland East of the Rocky Mountains: Guide for Selection of Practices for Soil and Water Conservation, Issues 282–290*; Agricultural Handbook 282; U.S. Department of Agriculture: Brooksville, FL, USA, 1965; p. 47.
37. Haan, C.T.; Barfield, B.J.; Hayes, J.C. *Design Hydrology and Sedimentology for Small Catchments*; Academic Press: San Diego, CA, USA, 1994; p. 588.
38. Diodato, N.; Aronica, G. Finding simplicity in storm erosivity modelling. In *Storminess and Environmental Change: Climate Forcing and Response in the Mediterranean Region*; Diodato, N., Bellocchi, G., Eds.; Springer: Dordrecht, The Netherlands, 2014; pp. 53–64.
39. Kirkby, M.; Abrahart, R.; McMahon, M.; Shao, J.; Thornes, J. MEDALUS soil erosion models for global change. *Geomorphology* **1998**, *24*, 35–49. [CrossRef]
40. Aloia, A.; Maietta, S.; Moretta, F.; Romano, E.; Mugnani, M.; Positano, M.P.; Tedesco, A.; Banco, C.; Andreola, V.; Liguori, V.; et al. *Progetto di Piano Stralcio Erosione Costiera*; Ai sensi dell'art. 17 commi 1 e 6 della Legge 18 maggio 1989 n. 183 s.m.i. e Legge 4 dicembre 1993 n. 493; Autorità di Bacino Regionale Sinistra Sele: Salerno, Italy, 2005; p. 29. (In Italian)
41. Pettitt, A.N. A non-parametric approach to the change-point detection. *Appl. Stat.* **1979**, *28*, 126–135. [CrossRef]
42. Nasta, P.; Palladino, M.; Ursino, N.; Saracino, A.; Sommella, A.; Romano, N. Assessing long-term impact of land-use change on hydrological ecosystem functions in a Mediterranean upland agro-forestry catchment. *Sci. Total. Environ.* **2017**, *605*, 1070–1082. [CrossRef]
43. Rizzi, F. *Osservazioni Statistiche Sul Cilento*; Tipografia di Angelo Trani: Naples, Italy, 1809; p. 6. (In Italian)
44. Van Oost, K.; Quine, T.A.; Govers, G.; De Gryze, S.; Six, J.; Harden, J.W.; Ritchie, J.C.; Mccarty, G.W.; Heckrath, G.; Kosmas, C.; et al. The Impact of Agricultural Soil Erosion on the Global Carbon Cycle. *Science* **2007**, *318*, 626–629. [CrossRef]
45. Borrelli, P.; Modugno, S.; Panagos, P.; Marchetti, M.; Schütt, B.; Montanarella, L. Detection of harvested forest areas in Italy using Landsat imagery. *Appl. Geogr.* **2014**, *48*, 102–111. [CrossRef]
46. Alvioli, M.; Melillo, M.; Guzzetti, F.; Rossi, M.; Palazzi, E.; Von Hardenberg, J.; Brunetti, M.T.; Peruccacci, S. Implications of climate change on landslide hazard in Central Italy. *Sci. Total. Environ.* **2018**, *630*, 1528–1543. [CrossRef]
47. Piacentini, T.; Galli, A.; Marsala, V.; Miccadei, E. Analysis of Soil Erosion Induced by Heavy Rainfall: A Case Study from the NE Abruzzo Hills Area in Central Italy. *Water* **2018**, *10*, 1314. [CrossRef]

water

MDPI

Article

Raindrop Energy Impact on the Distribution Characteristics of Splash Aggregates of Cultivated Dark Loessial Cores

Yu Fu [1], Guanglu Li [1,2,*], Dong Wang [2], Tenghui Zheng [1] and Mingxi Yang [2]

1 Institute of Soil and Water Conservation, Northwest A&F University, No. 26 Xinong Road, Yangling 712100, China

2 College of Resources and Environment, Northwest A&F University, No. 3 Taicheng Road, Yangling 712100, China

* Correspondence: guangluli@nwsuaf.edu.cn; Tel.: +86-029-87082376

Received: 15 June 2019; Accepted: 18 July 2019; Published: 21 July 2019

Abstract: To determine the effect of different rainfall energy levels on the breakdown of soil aggregates, this study analyzed the soil splash erosion amounts and the distribution of particle sizes under six rainfall conditions (rainfall energy: 2.41×10^{-5}–22.4×10^{-5} J m^{-2} s^{-1} and 1.29×10^{-4} J m^{-2} s^{-1}) at five splash distances (from 0–10 cm to 40–50 cm). Cores of the size 10×20 cm of undisturbed cultivated dark loessial soil were selected in tree replicates as the research subject. The results indicated that splashed aggregates were distributed mainly at splash distances of 0–20 cm, which accounted for 66%–90% of the total splash erosion amount. The splash erosion amount significantly decreased exponentially with increasing splash distance for the same rainfall energy ($p < 0.01$). The splash erosion amount significantly increased in the power function relationship with increasing rainfall energy at the same splash distance ($p < 0.05$). A model was obtained to predict the splash erosion amount for rainfall energy and splash distance. The fractal dimension (D) of the aggregates showed a downward opening parabolic relationship with raindrop energy. The maximal value of the rainfall energy was 1.286×10^{-4} J m^{-2} s^{-1}, which broke the aggregates to the largest degree. Enrichment ratio (ER) values for fragments >2 mm were close to 0. A particle size of 0.25 mm was the critical particle level for splash erosion.

Keywords: raindrop energy; soil aggregate; splash distance; fractal dimension; Loess Plateau

1. Introduction

Soil erosion reduces land productivity and soil fertility, destroys farmland, exacerbates flood disasters, and results in soil environment deterioration, which affects land exploitation and the protection and utilization of soil and water resources [1,2]. Water erosion is the main type of soil erosion in the Loess Plateau. The aggregates caused by splash erosion are dispersed and broken, which is the initial stage of water erosion [1–4]. Raindrops fall from the air and impact the surface soil particles at a certain speed under the action of gravity. Some soil particles are separated from the soil and become loose particles, which is conducive to the formation and flow of surface runoff. At the same time, the process provides abundant loose particles for subsequent runoff transport [5,6]. Additionally, soil particle transport leads to reduced or blocked pores on the soil surface and reduced soil permeability due to soil crust formation [7–9].

At present, research on splash erosion can be divided into two methods: Natural rainfall and artificial simulated rainfall. The observation of natural rainfall requires a long timeframe, and many factors are difficult to control, which makes the collection of splash data difficult. Artificial simulated rainfall can make up for shortages of natural rainfall and improve the feasibility of testing and reliability of the data. Rainfall characteristics include raindrop diameter, rainfall intensity, raindrop kinetic energy, and so on. The effect of rainfall intensity on raindrop diameter and rainfall was studied by Yakubu et al. [10], who proposed that the relationship between raindrop diameter and rainfall intensity could be expressed as an exponential or power function relationship. Sajjadi and Mahmoodabadi [11] carried out raindrop splashing tests under two different rainfall intensities (57 mm h^{-1} and 80 mm h^{-1}). The results showed that the splash erosion amounts of fine particles with sizes <0.043 mm were greater than those of large particles under a higher rainfall intensity. Moreover, with increasing rainfall duration, the rate of the total splash erosion observed at the beginning of the rainfall reaches a peak value: The larger the rainfall intensity, the less time it takes to reach the peak value [12]. Ziadat and Taimeh [13] reported that soil erosion on cultivated land was primarily affected by moisture content, whereas on uncultivated land it was mostly affected by slope steepness. Lim et al. [14] suggested that a power function relationship existed between raindrop energy and rainfall intensity. Free [15] showed that the splash erosion amount was positively correlated with rainfall energy and that raindrop energy was an important index to evaluate the degree of soil erosion caused by rainfall [14]. Rainfall intensity, raindrop energy, and their variations in time and space have important effects on the prediction of rainfall erosion [16,17].

At present, most research on raindrop energy for splash erosion has used a raindrop device to simulate natural rainfall [18] and control the raindrop energy by adjusting the rainfall intensity or height [16,19–21]. However, the rainfall energy of natural rainfall simulated by a raindrop generator is instantaneous energy, which has a range of changes, and an increase or decrease in the energy has a significant impact on the amount of raindrop splash erosion. Therefore, a raindrop generator with constant raindrop energy should be used for experimental studies to ensure that the results of the study are more accurate. The splash collection device was mainly composed of an Ellison splashing pan [22], an Ellison splashing cup [23], and a Morgan splashing pan [24]. The improved device [20,25,26] could collect most soil splash erosion particles, but the horizontal spatial distribution of the splash amount and particle size have been less well studied. Furthermore, most previous studies have used dried soil [27] or disturbed soil [20,28], which destroyed the soil structure and reduced the reliability of the data.

The objectives of this study were to (i) clarify the effect of raindrop energy on the splash distance and particle size distribution of aggregate splash erosion, (ii) establish a prediction equation for splash erosion in this area, and (iii) predict whether a certain raindrop energy will break up the aggregates to a great extent by comparing the characteristic parameters of the aggregate fractions.

2. Materials and Methods

2.1. Sampling Site and Soil Properties

The sampling site is located in the Changwu agriculture ecological experimental station, Weibei Plateau, Xianyang, Shanxi province (107°40′59″ E, 35°14′27″ N), and has an annual average temperature of 9.1 °C and annual average precipitation of 500.0 mm. The zonal soil is dark loessial soil [29]. The soil is loose, and the permeability is good, resulting in a good "soil reservoir" effect. The international soil texture classification is loam soil according to the USDA (United States Department of Agriculture) particle size classification criteria. Twenty-one samples and 1000 g of scattered soil were collected from the top layer (0–20 cm) of cultivated land using a cutting ring (10 cm diameter × 5 cm height) and the diagonal method. Three of the samples were used to determine the soil bulk density and moisture content, and the remaining 18 samples were used for the raindrop splash experiment. After the scattered soil dried naturally, the bulk density, moisture content, soil organic matter, total nitrogen, total phosphorus, and mechanical components were determined using the cutting ring (determine the bulk density and moisture content), potassium dichromate external heating, Kjeldahl nitrogen, $HClO_4$–H_2SO_4, and Malvin laser particle sizer methods. The soil characteristics of the sampling sites are reported in Table 1.

Table 1. Physical and chemical properties of the soil.

Soil Type	Bulk Density/ ($g{\cdot}cm^{-3}$)	Moisture Content/%	Soil Organic Carbon/%	Total Nitrogen/ ($g{\cdot}kg^{-1}$)	Total Phosphorus/ ($g{\cdot}kg^{-1}$)	Particle Size Composition/%		
						Sand (2–0.02 mm)	Silt (0.02–0.002 mm)	Clay (<0.002 mm)
Dark loessial soil	1.1 ± 0.1	18.8 ± 1.0	1.6 ± 0.1	1.0 ± 0.1	0.7 ± 0.03	52.3 ± 0.5	37.4 ± 0.03	10.3 ± 0.04
	Fragment size distribution of the undisturbed soil/%							
	>2 mm	1–2 mm	0.5–1 mm	0.25–0.5 mm		0.053–0.25 mm		<0.053 mm
	6.9 ± 1.8	12.3 ± 1.1	21.3 ± 4.2	21.1 ± 2.1		14.6 ± 4.1		23.9 ± 2.9

2.2. Experimental Design

The artificial rainfall device used in this study consisted of two parts: A raindrop generator and a splashed raindrop-collecting device (Figure 1) [30]. The raindrop generator was a cylindrical box with an open top (10 cm in diameter and 10 cm in height). Twenty-one syringe needles were installed in the floor of the box, and different needles could generate different raindrop diameters. A stainless steel pan (110 cm diameter) containing six concentric circles composed of wire was selected as the splash raindrop-collecting device. The splash pan was centered in a circle and was used to place the open cutting ring (10 cm diameter × 5 cm height). In turn, the edge of the cutting ring contained fenced concentric circles with wire at distances of 10 cm, 20 cm, 30 cm, 40 cm, and 50 cm. The area between the splash pan and wire was impermeable. Each concentric circle was set symmetrically with two drains, which were used to collect splashed soil particles. Outside of the experimental device, a baffle and plastic cover were placed to prevent the effects of horizontal airflow disturbance on the rainfall.

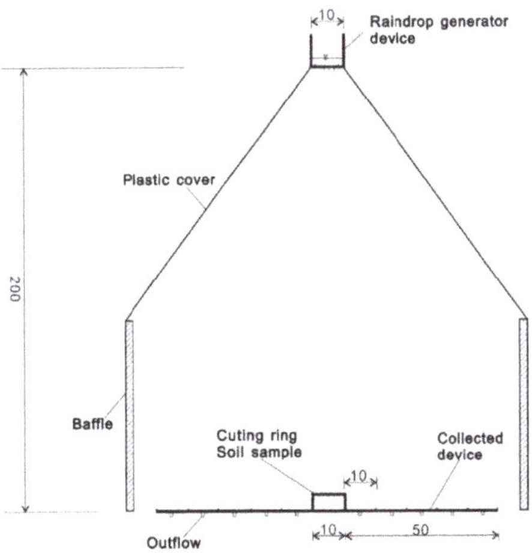

Figure 1. The test for splashed raindrops and soil aggregates (unit: cm). Note: the figure is from Fu et al. [30].

Six raindrop energy levels determined by six raindrop diameters and the height of the rainfall (2 m) were selected in the experiment. The corresponding six raindrop diameters were 2.67 mm, 3.05 mm, 3.39 mm, 3.79 mm, 4.05 mm, and 5.45 mm. The raindrop energy was calculated by the following method: When the raindrop diameter was greater than or equal to 1.9 mm, the final velocity of the raindrops was calculated using a modified Newton formula (Equation (1)). Equation (2) was used to calculate the raindrop velocity under this test condition [31]. The raindrop energy was calculated by Equation (3). Therefore, the final raindrop energy levels were 2.41×10^{-5}, 3.68×10^{-5}, 5.15×10^{-5}, 7.30×10^{-5}, 8.97×10^{-5}, and 22.4×10^{-5} J m^{-2} s^{-1}, which conformed to the range of natural rainfall raindrop energy. The corresponding rainfall intensities and rainfall diameters all conformed to the rainfall characteristics of the Loess Plateau [32]. Equations (1)–(3) are

$$V_i = (17.20 - 0.84d)\sqrt{0.1d},\ \ d > 1.9, \tag{1}$$

$$V = V_i\sqrt{1 - e^{-\frac{2g}{V_i^2}H}}, \tag{2}$$

$$E_{rs} = \sum_{i=1}^{n} \frac{1}{2}\, \mathrm{m}V^2, \tag{3}$$

where V is the raindrop velocity (m s^{-1}), d is the raindrop diameter (mm), V_i is the terminal velocity (m s^{-1}), H is the height of the falling raindrop (m), E_{rs} is the raindrop energy (J m^{-2} s^{-1}), m is the individual raindrop mass (g), $i = 0, \ldots, n$ is the number of raindrops, and g is gravity acceleration (m s^{-2}).

Prior to the start of the test, the water head height was adjusted to 2 cm. All of the needles started to drop raindrops at the same time. The rainfall duration was 10 min, and all tests were replicated 3 times. The splashed fragments were collected for each of the five distances (0–10 cm, 10–20 cm, 20–30 cm, 30–40 cm, and 40–50 cm). An aggregate analyzer (HR-TTF-100, Shunlong Experiment Instrument Factory, Yuxi City, Zhejiang province, China) was used to sieve the fragments into size fractions of >2 mm, 1–2 mm, 0.5–1 mm, 0.25–0.5 mm, 0.053–0.25 mm, and <0.053 mm, which then were oven-dried for 24 h at 105 °C and weighed. A new test plot was prepared after each rainfall event.

2.3. Parameter Calculation

This study used the fractal dimension (D) proposed by Yang [33] with the quality of the particle size distribution described in the soil fractal model as follows:

$$\frac{W(\delta > d_i)}{W_0} = 1 - \left(\frac{d_i}{d_{max}}\right)^{3-D}. \tag{4}$$

Take the exponential of both sides,

$$\frac{W(\delta > d_i)}{W_0} = (3 - D)\lg\left(\frac{d_i}{d_{max}}\right), \tag{5}$$

where $W(\delta > d_i)$ is the cumulative mass of soil particles with size δ more than a comparative size d_i, W_0 is the total mass of the particles, d_i is the average of the sieve size range (dj, dj + 1), and d_{max} is the maximum particle size.

For each size fraction, the enrichment ratio (ER) refers to the mass percentage of splash sediment compared to the undisturbed aggregates. An ER value >1 indicates enrichment of the fraction, whereas an ER value < 1 demonstrates that the fraction is depleted. The ER is calculated as

$$ER = \frac{P_{sp}}{P_{sa}}, \tag{6}$$

where P_{sp} is the mass percentage of one size fraction in the total mass of splash aggregates, and P_{sa} is the mass percentage of one size fraction in the total mass of undisturbed aggregates.

Nash–Sutcliffe indices (E_{NS}) [34] were used to evaluate the model prediction, which is a modeling efficiency metric ranging from negative infinity to 1, where only >0 values are acceptable:

$$E_{NS} = 1 - \frac{\sum_{i=1}^{n}(O_i - S_i)^2}{\sum_{i=1}^{n}\left(O_i - \overline{O}\right)^2}, \tag{7}$$

where O_i is the ith measured value of the mass of splash erosion, S_i is the ith predicted value of the mass of splash erosion, n is the total number of measured values, and $\overline{O}$ is the mean of the measured values of the mass of splash erosion. Significant differences in the splash erosion amounts for different raindrop energies or different splash distances were detected using one-way analysis of variance (ANOVA), followed by the LSD test (Least–Significant Difference) ($p < 0.05$). The relationships between the splash erosion amount and splash distances, between the splash erosion amount and raindrop energy, and between the splash erosion amounts, splash distances, and raindrop energy levels were analyzed using a simple regression method. All statistical analyses were performed using SPSS 16.0 (IBM SPSS Software, Armonk, NY, USA), and all figures were processed in Origin 8.5 (OriginLab Corporation, Northampton, MA, USA).

3. Results and Discussion

3.1. Distribution Characteristics of the Splash Erosion Amounts

The distribution of the splash erosion amounts under the different raindrop energy levels is shown in Table 2. The splash erosion amount decreased with increasing splash distance for the same raindrop energy. The splash erosion amounts of the six raindrop conditions were distributed mainly at 0–20 cm, which accounted for 66% of the total splash erosion amount and was significantly higher than the splash erosion amounts at the other distances. The splash erosion amounts at 20–30 cm and 30–40 cm accounted for 16% and 10% of the total splash erosion amount, respectively. In addition, the amount of splash erosion that reached a distance of 40 to 50 cm accounted for 1%–8% of the total splash erosion amount, which was significantly less than for the other splash distances ($p < 0.05$). The raindrops collided with the surface soil to destroy the original soil structure, which caused them to disperse and spatter soil particles under the action of their own gravity via parabolic motion. Their movement paths were affected by their own weight and volume. Theoretically, when raindrop energy is the same, small-sized particles are transported longer distances because of their light mass, whereas large-sized particles remain at close distances [29]. Therefore, the amount of aggregate fractions at the 0–20 cm distance was significantly higher than that at the other splash distances. Regression analysis of the splash erosion amount (M) and splash distance (S) produced for the different raindrop energy levels showed that the splash amount M decreased exponentially with increasing splash distance S ($p < 0.01$). The regression equation was $M = m exp(nS)$, where m and n are the parameters (Table 3). This result was consistent with those of the studies of Dijk et al. [26], Cheng et al. [12], and Fu et al. [30].

Table 2. Amounts of splash erosion at different distances for different raindrop energy levels.

Raindrop Energy/(J m^{-2} s^{-1})	Splash Erosion Amount of Different Splash Distances/g					Total Splash Erosion Amount/g
	0–10 cm	10–20 cm	20–30 cm	30–40 cm	40–50 cm	
2.41×10^{-5}	1.1 ± 0.2 Aa	0.9 ± 0.2 Aa	0.5 ± 0.2 Ba	0.3 ± 0.2 Ba	0.3 ± 0.1 Ba	3.0 ± 0.7
3.68×10^{-5}	3.2 ± 1.0 Aa	2.0 ± 0.7 Ba	0.8 ± 0.3 Ca	0.4 ± 0.1 Ca	0.2 ± 0.1 Ca	6.5 ± 2.2
5.15×10^{-5}	7.5 ± 0.2 Ab	2.7 ± 0.1 Ba	0.9 ± 0.1 Ca	0.4 ± 0.03 Da	0.2 ± 0.02 Ea	11.7 ± 0.4
7.30×10^{-5}	26.4 ± 2.5 Ac	6.3 ± 0.8 Bb	2.2 ± 0.2 Cb	1.0 ± 0.1 Cb	0.4 ± 0.1 Ca	36.4 ± 2.3
8.97×10^{-5}	28.1 ± 2.3 Acd	7.9 ± 1.3 Bb	2.7 ± 0.5 Cb	1.3 ± 0.2 Cb	0.5 ± 0.1 Ca	40.4 ± 4.3
22.4×10^{-5}	30.0 ± 2.9 Ad	17.8 ± 3.0 Bc	7.2 ± 1.3 Cc	3.5 ± 0.5 Dc	1.6 ± 0.4 Db	60.1 ± 3.2

Note: The lowercase letters represent the difference in the amount of splash erosion between the same splash distance and different raindrop energy levels, and the uppercase letters denote the difference in splash amount between the same raindrop energy level and different splash distances. There was no significant difference in the values of each column or line with the same letter ($p > 0.05$).

Table 3. Relationships between the amounts of splash (M) and the splash distance (S) for different raindrop energy levels.

Raindrop Energy/(J m^{-2} s^{-1})	Simulated Equation	R^2	p-Value	n
2.41×10^{-5}	$M = 1.725 \times e^{-0.040S}$	0.976	<0.01	30
3.68×10^{-5}	$M = 6.826 \times e^{-0.0471S}$	0.991	<0.01	30
5.15×10^{-5}	$M = 16.553 \times e^{-0.091S}$	0.990	<0.01	30
7.30×10^{-5}	$M = 57.291 \times e^{-0.101S}$	0.985	<0.01	30
8.97×10^{-5}	$M = 61.771 \times e^{-0.098S}$	0.989	<0.01	30
22.4×10^{-5}	$M = 68.814 \times e^{-0.074S}$	0.995	<0.01	30

When the splash distance was the same (Table 2), the splash erosion amount increased with increasing raindrop energy. The splash erosion amount of the raindrop energy 22.4×10^{-5} J m^{-2} s^{-1} was at a maximum and was significantly higher than that of the other raindrop energies, which were 1.1–27.3, 2.3–20.0, 2.7–15.0, 2.7–10.9, and 3.1–6.7 times that of the other splash distances, respectively. The regression analysis showed that for the same splash distance, the splash erosion amount M had a power function relationship with the raindrop energy E (Table 4). These results were consistent with those of Hu et al. [20], Rose [35], Parsons et al. [36], and Fernándezraga et al. [37] and showed that the impact of the raindrop was enhanced by an increase in the raindrop energy, which then promoted the separation and transportation of more soil particles. The index value of the relationship between the total splash erosion amount and the raindrop energy was 1.42. This value was close to the index value of 1.46 from the study of Free [15]. However, Hu et al. [20] showed that the total soil splash erosion amount was a power function with increasing raindrop energy in the black soil area of the northeast Chinese mollisol region under rainfall intensities of 50 mm h^{-1} and 100 mm h^{-1}: Moreover, the index values were 2.79 and 1.69, respectively, which were different from the results of this study. This discrepancy may have been due to differences in the soil organic matter content between black loessial soil (SOM (Soil Organic Matter): 1.55%) and the black soil area (SOM: 2.38%), because the soil organic matter content may have affected soil erodibility. The relationship between the splash erosion amount and raindrop energy varied with increasing splash distances, and both the raindrop energy (E) and splash distance (S) affected the splash erosion amount (M). For the M values, E values, and splash distances S that were analyzed using multiple regression, the relationship followed a power–exponential function: $M = E^{0.72} \times 16.28e^{12.92 \times S^{-0.21}}$, $R^2 = 0.81$, $p < 0.05$. The data of Wang et al. [21] were placed into the above equation for verification. The E_{NS} value was 0.67 (Figure 2a). It can be seen that the fit was better [34], which may have been because the soil used in Wang et al. [21] was from the Loess Plateau and the rainfall conditions were similar to this study. The data of Cheng et al. [12] were placed into the above equation for verification. The E_{NS} value was less than 0 (Figure 2b). This may have been affected by soil types and slope. The data in this study were placed into the splash erosion model of Hu et al. [20]: $S_T = 0.14KE^{2.65}D_{50}{}^{0.54}$, where S_T is total splash erosion (g), KE is raindrop kinetic energy (J m^{-2} mm^{-1}), and D_{50} is the raindrop median volume diameter (mm). The results showed that the E_{NS} value was less than 0 (Figure 3), so the model of Hu et al. [20] could not predict the amount of splash erosion in this study, which was mainly affected by soil types and rainfall conditions. The above research further indicated that the model proposed in this study can evaluate soil splash erosion in the Loess Plateau well, but the simulation results of the areas outside the Loess Plateau are still insufficient. Therefore, model correction needs further exploration in future research.

Table 4. Relationships between the splash erosion amount (M) and raindrop energy (E) at different splash distances.

Splash Distance	Simulated Equation	R^2	p-Value	n
0–10 cm	$M = 3.332 \times 10^7 E^{1.564}$	0.779	<0.05	18
10–20 cm	$M = 2.144 \times 10^6 E^{1.365}$	0.964	<0.05	18
20–30 cm	$M = 3.325 \times 10^5 E^{1.217}$	0.972	<0.05	18
30–40 cm	$M = 5.663 \times 10^4 E^{1.159}$	0.942	<0.05	18
40–50 cm	$M = 3.511 \times 10^3 E^{0.941}$	0.837	<0.05	18
0–50 cm	$M = 1.515 \times 10^7 E^{1.421}$	0.872	<0.05	18

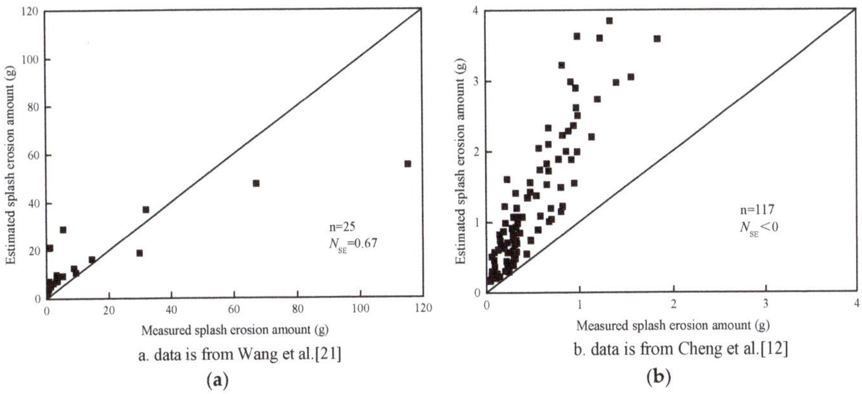

Figure 2. Comparison between the predicted and measured values of the splash erosion amount. (**a**) Data are from Wang et al. [21]; (**b**) data are from Cheng et al. [12].

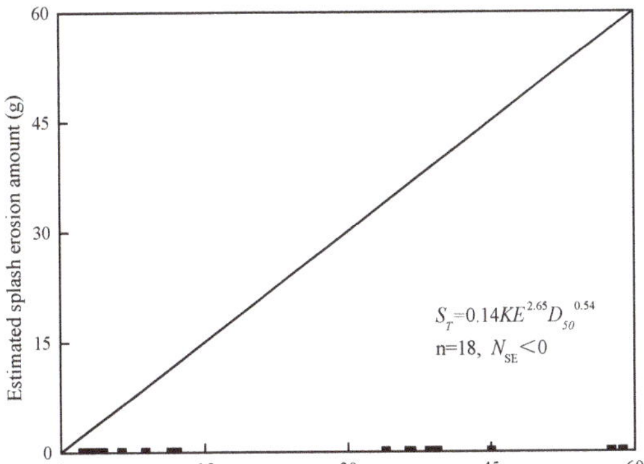

Figure 3. Comparison between the predicted and measured values of the splash erosion amount.

3.2. Particle Size Distribution Characteristics of Splash Aggregate Fragments

The particle size distribution of splashed aggregates for different raindrop energy levels is shown in Figure 4. For each raindrop energy level, the mass percentage of soil aggregates >2 mm was almost zero, possibly because these aggregates were not sufficient to splash out under the present experimental conditions [25,30]. As a whole, the mass percentage of particle sizes >0.25 mm presented a down up trend with increasing raindrop energy. When the raindrop energy was 2.41×10^{-5} J m^{-2} s^{-1}, the mass percentage of the water-stable aggregates >0.25 mm was highest, with a value of 39%. When the raindrop energy was 7.30×10^{-5} J m^{-2} s^{-1}, the mass percentage of water-stable aggregates >0.25 mm in diameter was lowest, with a value of 18%. This outcome may have been due to the smaller raindrop energy condition, in which the raindrop was less effective at soil splash erosion. With an increase in the raindrop energy, the effect of raindrops on soil aggregates was enhanced. Therefore, dispersed soil particles were more likely to be splashed, and some of the splashed aggregates broke into microaggregates (<0.25 mm). Due to the lighter mass of the microaggregates themselves, the quantity of transportation was larger. When the raindrop energy was higher than 7.30×10^{-5} J m^{-2} s^{-1}, more macroaggregates splashed out, and thus the mass percentage of the macroaggregates (>0.25 mm) increased.

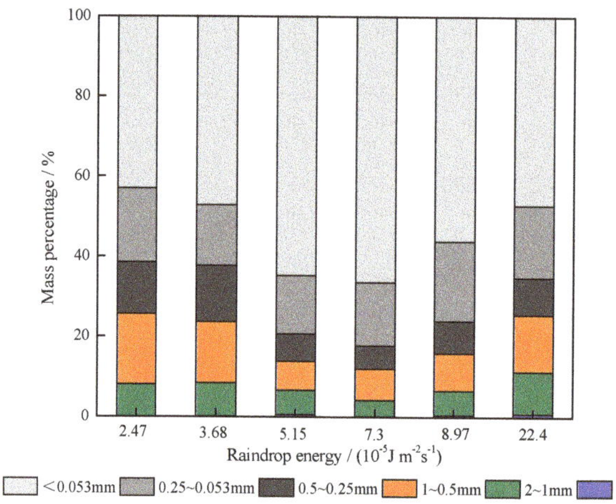

Figure 4. The particle size distribution of splashed aggregates for different raindrop energy levels.

3.3. *Effect of Raindrop Energy Levels on Characteristic Parameters of Splash Aggregate Fragments*

Castrignano [38] reported that larger D values resulted in more dispersion of aggregates. Lower D values of aggregates represented a particle size distribution dominated by larger fragments, whereas higher D values reflected a distribution dominated by smaller fragments [39]. The fractal dimensions of the splash aggregate fragments for the six raindrop energy levels were 2.789, 2.798, 2.896, 2.898, 2.862, and 2.818, respectively. Regression analysis of the fractal dimension and raindrop energy levels concluded that the fractal dimension of the soil splash aggregate (D) clearly changed as a quadratic function, which presented an up–down trend with increasing raindrop energy (E). The equation is $D = -1.0532 \times 10^{-4} (E - 1.29 \times 10^{-4})^2 + 2.564$, $R^2 = 0.627$, $p < 0.05$. The maximal raindrop energy value was 1.29×10^{-4} J m^{-2} s^{-1}, which broke the aggregates to the largest degree. On the whole, the fractal dimension of the splash aggregate fragments for the raindrop energy of 2.41×10^{-5} J m^{-2} s^{-1} was lower than that for 22.4×10^{-5} J m^{-2} s^{-1}. This difference may have been due to the lowest raindrop energy level, less splashing of aggregates, and a higher proportion of large particles splashed out due to their own mass. However, when the raindrop energy was larger, more macroaggregates splashed out, although some of them broke into microaggregates [28]: Therefore, the fractal dimension was relatively high.

The splash enrichment ratio (ER) for each particle size at the different raindrop energy levels is shown in Figure 5. The enrichment ratio of aggregates presented an up–down trend with decreasing particle size. The ER value for a particle size greater than 2 mm was close to 0. Ma et al. [25] showed that a rainfall intensity of 58.1 mm·h^{-1} resulted in ER values of 0 for particle sizes of 5–2 mm and a value close to 0 for particle sizes of 2–1 mm, which was consistent with the results of this study. The ER value for particle sizes greater than 0.25 mm was <1, whereas that for particle sizes less than 0.25 mm was >1 for all raindrop energy levels. Thus, the 0.25-mm particle size was the critical particle size level for the enrichment and depletion of splash erosion. However, Zhou [40] analyzed the selective characteristics of splash erosion for black soil aggregates and concluded that aggregates with a particle size >1.0 mm were not easily transported. Hence, a particle size of 1.0 mm was the critical particle level for the enrichment and depletion of splash erosion. This finding was inconsistent with the conclusion of this study, possibly because the soil selected in this study was developed from loess parent material, and the erodibility of the soil was relatively high. Under the condition of raindrop splash erosion, the large particle size was smaller and was easily broken into smaller particle-sized aggregates: Thus, the >0.25-mm macroaggregates were gradually broken into <0.25-mm microaggregates. This result was consistent with that of Legout et al. [28], who proposed that soil aggregates could be hypothesized into a single stratified structure. The macroaggregates (>0.25 mm) were bound by microaggregates (<0.25 mm) and gradually disintegrated into microaggregates during rainfall. The macroaggregates were mainly broken into 0.25–0.053-mm aggregates for all rainfall experiments (Figure 5). Fu et al. [41] considered that the *ER* value of particle sizes 1–0.053 mm was greater than 1. Aggregates with a particle size >1 mm were mainly broken into aggregates with particle sizes of 1–0.053 mm in the single-raindrop test in that study. That particle size was larger than the results of this study, because the single-raindrop splash soil resulted in less fragmentation of the aggregates [41]. Conversely, in a study with multineedle rainfall, soil particles may receive secondary raindrop splash erosion, and hence more fine particles may be produced.

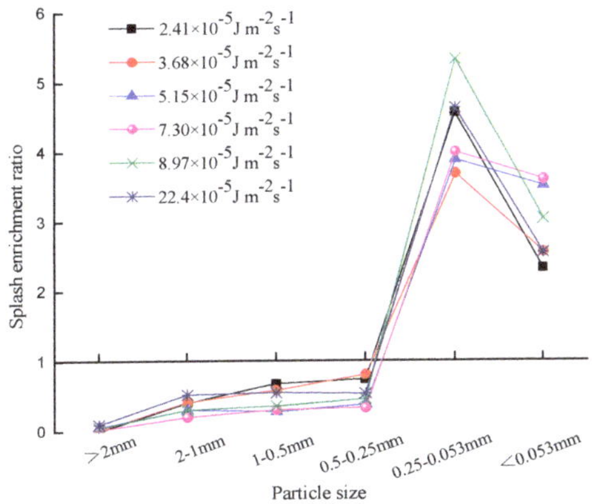

Figure 5. Splash enrichment ratio (*ER*) for each particle size of the different raindrop energy levels.

4. Conclusions

This study analyzed the soil splash erosion amounts and the distribution of particle sizes under six rainfall conditions at five splash distances. Dark loessial soil with a Corg content (Organic Carbon) of 1.6% and natural moisture of 18.8% and a core size of 10 × 20 cm was used in laboratory experiments.

The splash erosion amounts of the six raindrop conditions were mainly distributed at distances of 0–20 cm. For each raindrop energy level, the mass percentage of soil aggregates >2 mm was almost zero. As a whole, the mass percentage of particle sizes >0.25 mm presented a down–up trend. A particle size of 0.25 mm was the critical particle size level for the enrichment and depletion of splash erosion.

A model was obtained to predict the splash erosion amount for the rainfall energy and splash distance, which could evaluate the soil splash erosion in the Loess Plateau well, but the simulation results of the areas outside the Loess Plateau are still insufficient. Therefore, model correction needs further exploration in future research.

The fractal dimension (*D*) of the aggregates showed a downward opening parabolic relationship with raindrop energy. The maximal raindrop energy value was 1.29×10^{-4} J m^{-2} s^{-1}, which broke the aggregates to the largest degree. The above results provide insights into the variation in aggregates in the soil layer during rainfall and a good understanding of soil surface crust and soil erosion mechanisms.

Author Contributions: Conceptualization, Y.F., G.L., and D.W.; data curation, Y.F.; formal analysis, Y.F., G.L., and D.W.; funding acquisition, G.L.; investigation, Y.F., G.L., D.W., and T.Z.; methodology, Y.F.; project administration, G.L.; resources, D.W.; software, Y.F. and G.L.; supervision, G.L.; validation, Y.F. and G.L.; writing—original draft, Y.F.; writing—review and editing, Y.F., G.L., D.W., T.Z. and M.Y.

Funding: This work was supported by the National Natural Science Foundation of China (41571262) and the Chinese Ministry of Water Resources Science and Technology Promotion (TG1308).

Acknowledgments: The authors thank anonymous reviewers for insightful comments on the original manuscript. This work was supported by the National Natural Science Foundation of China (Grant. No. 41571262) and the Chinese Ministry of Water Resources Science and Technology Promotion program (TG1308).

Conflicts of Interest: The authors declare no conflict of interest.

References

1. Kinnell, P.I.A. Raindrop impact induced erosion processes and prediction: A review. *Hydrol. Process.* **2005**, *19*, 2815–2844. [CrossRef]
2. Hu, F.N.; Liu, J.F.; Xu, C.Y.; Wang, Z.L.; Liu, G.; Li, H.; Zhao, S.W. Soil internal forces initiate aggregate breakdown and splash erosion. *Geoderma* **2018**, *320*, 43–51. [CrossRef]
3. Li, G.L.; Fu, Y.; Li, B.Q.; Zheng, T.H.; Wu, F.Q.; Peng, G.Y.; Xiao, T.Q. Micro–characteristics of soil aggregate breakdown under raindrop action. *Catena* **2017**, *162*, 354–359. [CrossRef]
4. Xiao, H.; Liu, G.; Abd-Elbasit, M.A.M.; Zhang, X.C. Effects of slaking and mechanical breakdown on disaggregation and splash erosion. *Eur. J. Soil Sci.* **2017**, *68*, 797–805. [CrossRef]
5. Wuddivira, M.N.; Stone, R.J.; Ekwue, E.I. Clay, Organic Matter, and Wetting Effects on Splash Detachment and Aggregate Breakdown under Intense Rainfall. *Soil Sci. Soc. Am. J.* **2009**, *73*, 226–232. [CrossRef]
6. Fernández–Raga, M.; Palencia, C.; Keesstra, S.; Jordán, A.; Fraile, R.; Angulo–Martínez, M.; Cerdà, A. Splash erosion: A review with unanswered questions. *Earth–Sci. Rev.* **2017**, *171*, 463–477. [CrossRef]
7. Ramos, M.C.; Nacci, S.; Pla, I. Effect of raindrop impact and its relationship with aggregate stability to different disaggregation forces. *Catena* **2003**, *53*, 365–376. [CrossRef]
8. Saedi, T.; Shorafa, M.; Gorji, M.; Khalili, M.B. Indirect and direct effects of soil properties on soil splash erosion rate in calcareous soils of the central Zagross, Iran: A laboratory study. *Geoderma* **2016**, *271*, 1–9. [CrossRef]
9. Lado, M.; Benhur, M.; Shainberg, I. Soil wetting and texture effects on aggregate stability, seal formation, and erosion. *Soil Sci. Soc. Am. J.* **2004**, *68*, 1992–1999. [CrossRef]
10. Yakubu, M.L.; Yusop, Z.; Fulazzaky, M.A. The influence of rain intensity on raindrop diameter and the kinetics of tropical rainfall: Case study of Skudai, Malaysia. *Int. Assoc. Sci. Hydrol. Bull.* **2016**, *61*, 944–951. [CrossRef]
11. Sajjadi, A.S.; Mahmoodabadi, M. Aggregate breakdown and surface seal development influenced by rain intensity, slope gradient and soil particle size. *Solid Earth* **2015**, *6*, 311–321. [CrossRef]
12. Cheng, J.; Qin, Y.; Zhang, H.; Cong, Y.; Yang, F.; Yan, Y.Q. Splash erosion under artificial rainfall in Rocky mountain area of northern China. *Trans. Chin. Soc. Agric. Mach.* **2015**, *46*, 153–161.
13. Ziadat, F.M.; Taimeh, A.Y. Effect of rainfall intensity, slope, land use and antecedent soil moisture on soil erosion in an arid environment. *Land Degrad. Dev.* **2013**, *24*, 582–590. [CrossRef]
14. Lim, Y.S.; Jin, K.K.; Kim, J.W.; Park, B.I.; Kim, M.S. Analysis of the relationship between the kinetic energy and intensity of rainfall in Daejeon, Korea. *Quatern. Int.* **2015**, *384*, 107–117. [CrossRef]
15. Free, G.R. Erosion characteristics of rainfall. *Agric. Eng.* **1960**, *41*, 447–449.
16. Dijk, A.I.J.M.V.; Bruijnzeel, L.A.; Rosewell, C.J. Rainfall intensity–kinetic energy relationships: A critical literature appraisal. *J. Hydrol.* **2002**, *261*, 1–23. [CrossRef]
17. Angulo–Martínez, M.; Beguería, S.; Navas, A.; Machín, J. Splash erosion under natural rainfall on three soil types in NE Spain. *Geomorphology* **2012**, *175*, 38–44. [CrossRef]
18. Dai, X.; Zhang, H.T.; Zhou, D.M. Introduction to artificial rainfall simulator and its application. *Soil Water Conserv. China* **2012**, *12*, 52–54.

19. Shainberg, I.; Mamedov, A.I.; Levy, G.J. Role of wetting rate and rain energy in seal formation and erosion 1. *Soil Sci.* **2003**, *168*, 54–62. [CrossRef]
20. Hu, W.; Zheng, F.; Bian, F. The Directional Components of Splash Erosion at Different Raindrop Kinetic Energy in the Chinese Mollisol Region. *Soil Sci. Soc. Am. J.* **2016**, *80*, 38–44. [CrossRef]
21. Wang, D.; Li, G.L.; Fu, Y.; Gao, G.X.; Zheng, T.H. Detachment and transport characteristics of sandy loam soil by raindrop action in the northern loess plateau, China. *J. Soil Water Conserv.* **2018**, *73*, 705–713. [CrossRef]
22. Ellison, W.D. Studies of raindrop erosion. *Agric. Eng.* **1944**, *25*, 181–182.
23. Ellison, W.D. Two devices for measuring soil erosion. *Agric. Eng.* **1944**, *25*, 53–55.
24. Morgan, R.P.C. Field studies of rainsplash erosion. *Earth Surf. Proc. Land.* **1978**, *3*, 295–299. [CrossRef]
25. Ma, R.M.; Li, Z.X.; Cai, C.F.; Wang, J.G. The dynamic response of splash erosion to aggregate mechanical breakdown through rainfall simulation events in Ultisols (subtropical China). *Catena* **2014**, *121*, 279–287. [CrossRef]
26. Dijk, A.I.J.M.V.; Meesters, A.G.C.A.; Bruijnzeel, L.A. Exponential distribution theory and the interpretation of splash detachment and transport experiments. *Soil Sci. Soc. Am. J.* **2002**, *66*, 1466–1474. [CrossRef]
27. Ellison, W.D. Soil erosion studies. Parts I and II. *Agric. Eng.* **1947**, *28*, 197–201.
28. Legout, C.; Leguédois, S.; Le Bissonnais, Y. Aggregate breakdown dynamics under rainfall compared with aggregate stability measurements. *Eur. J. Soil Sci.* **2005**, *56*, 225–238. [CrossRef]
29. The Ministry of Water Resources of the People's Republic of China; Chinese Academy of Sciences; Chinese Academy of Engineering. *Water Loss and Soil Erosion and Ecological Security of China: The Loess Plateau*; Science Press: Beijing, China, 2010.
30. Fu, Y.; Li, G.L.; Zheng, T.H.; Zhang, T. Splash detachment and transport of loess aggregate fragments by raindrop action. *Catena* **2017**, *150*, 154–160. [CrossRef]
31. Yao, W.Y.; Chen, G.X. Calculation formula of rain drop fall velocity. *J. Hohai Univ.* **1993**, *21*, 21–27.
32. Jiao, J.Y.; Wang, W.Z.; He, X.P. Precipitation and erosion characteristics of rain–storm in different pattern on Loess Plateau. *J. Arid Land Resour. Environ.* **1999**, *13*, 34–42.
33. Yang, P.L.; Luo, Y.P.; Shi, Y.C. Using weight distribution of soil particle size to express soil fractal features. *Chin. Sci. Bull.* **1993**, *38*, 1896–1899.
34. Nash, J.E.; Sutcliffe, J.V. River flow forecasting through conceptual models 1: A discussion of principles. *J. Hydrol.* **1970**, *10*, 282–290. [CrossRef]
35. Rose, C.W. Soil detachment caused by rainfall. *Soil Sci.* **1960**, *89*, 28–35. [CrossRef]
36. Parsons, A.J.; Abrahams, A.D.; Wainwright, J. Rainsplash and erosion rates in an interrill area on semi–arid grassland southern Arizona. *Catena* **1994**, *22*, 215–226. [CrossRef]
37. Fernándezraga, M.; Fraile, R.; Keizer, J.J.; Teijeiro, M.E.V.T.; Castro, A.; Palencia, C.; Calvo, A.I.; Koenders, J.; Marques, R.L.D.C. The kinetic energy of rain measured with an optical disdrometer: An application to splash erosion. *Atmos. Res.* **2010**, *96*, 225–240. [CrossRef]
38. Castrignanò, A.; Stelluti, M. Fractal Geometry and Geostatistics for describing the Field Variability of Soil Aggregation. *J. Agric. Eng. Res.* **1999**, *73*, 13–18. [CrossRef]
39. Tyler, S.W.; Wheatcraft, S.W. Fractal scaling of soil particle–size distributions: Analysis and limitations. *Soil Sci. Soc. Am. J.* **1992**, *56*, 362–369. [CrossRef]
40. Zhou, Y.Y.; Wang, E.H.; Chen, X.W. Splash erosion of black soil with different reclamation year sand its relations to soil aggregates selective characteristic. *Chin. J. Appl. Ecol.* **2009**, *20*, 2411–2416.
41. Fu, Y.; Li, G.L.; Zheng, T.H.; Zhang, T. Effects of raindrop splash on aggregate particle size distribution of soil plough layer. *Trans. Chin. Soc. Agric. Eng.* **2017**, *33*, 155–160.

water

MDPI

Article

Projected Rainfall Erosivity Over Central Asia Based on CMIP5 Climate Models

Eldiiar Duulatov [1,2,3], Xi Chen [1,2,*], Amobichukwu C. Amanambu [1,2,4], Friday U. Ochege [1,2,5], Rustam Orozbaev [3,6], Gulnura Issanova [7,8,9] and Gulkaiyr Omurakunova [1,2]

1 State Key Laboratory of Desert and Oasis Ecology, Xinjiang Institute of Ecology and Geography, Chinese Academy of Sciences, Urumqi 830011, China; e.duulatov89@gmail.com (E.D.); a.amanambu@ufl.edu (A.C.A.); uchenna.ochege@uniport.edu.ng (F.U.O.); omurakunova1989@mail.ru (G.O.)
2 University of Chinese Academy of Science, Beijing 100049, China
3 Institute of Geology, National Academy of Sciences of Kyrgyz Republic, 30 Erkindik, Bishkek 720040, Kyrgyzstan; r.t.orozbaev@gmail.com
4 Department of Geography, University of Florida, Gainesville, FL 32611-7315, USA
5 Department of Geography and Environmental Management, University of Port Harcourt, East-West Rd., PMB 5323 Choba, Port Harcourt, Rivers State, Nigeria
6 Applied Geology Department, American University of Central Asia, 7/6 Aaly Tokombaev, Bishkek 720060, Kyrgyzstan
7 Faculty of Geography and Environmental Sciences, Al-Farabi Kazakh National University, Ave. Al-Farabi 71, Almaty 050040, Kazakhstan; agamprit@gmail.com
8 Institute of Natural Sciences and Geography, Abai Kazakh National Pedagogical University, Ave. Dostyk 13, Almaty 050010, Kazakhstan
9 Research Centre for Ecology and Environment of Central Asia (Almaty), Ave. Al-Farabi 75 V, Almaty 050060, Kazakhstan
* Correspondence: chenxi@ms.xjb.ac.cn; Tel.: +86-136-0992-3012

Received: 23 February 2019; Accepted: 24 April 2019; Published: 28 April 2019

check for updates

Abstract: Climate change-induced precipitation variability is the leading cause of rainfall erosivity that leads to excessive soil losses in most countries of the world. In this paper, four global climate models (GCMs) were used to characterize the spatiotemporal prediction of rainfall erosivity and assess the effect of variations of rainfall erosivity in Central Asia. The GCMs (BCCCSM1-1, IPSLCM5BLR, MIROC5, and MPIESMLR) were statistically downscaled using the delta method under Representative Concentration Pathways (RCPs) 2.6 and 8.5 for two time periods: "Near" and "Far" future (2030s and 2070s). These GCMs data were used to estimate rainfall erosivity and its projected changes over Central Asia. WorldClim data was used as the present baseline precipitation scenario for the study area. The rainfall erosivity (*R*) factor of the Revised Universal Soil Loss Equation (RUSLE) was used to determine rainfall erosivity. The results show an increase in the future periods of the annual rainfall erosivity compared to the baseline. For all GCMs, with an average change in rainfall erosivity of about 5.6% (424.49 MJ mm ha^{-1} h^{-1} year^{-1}) in 2030s and 9.6% (440.57 MJ mm ha^{-1} h^{-1} year^{-1}) in 2070s as compared to the baseline of 402 MJ mm ha^{-1} h^{-1} year^{-1}. The magnitude of the change varies with the GCMs, with the largest change being 26.6% (508.85 MJ mm ha^{-1} h^{-1} year^{-1}), occurring in the MIROC-5 RCP8.5 scenario in the 2070s. Although annual rainfall erosivity shows a steady increase, IPSLCM5ALR (both RCPs and periods) shows a decrease in the average erosivity. Higher rainfall amounts were the prime causes of increasing spatial-temporal rainfall erosivity.

Keywords: rainfall erosivity; Central Asia; GCMs; soil erosion; climate change

1. Introduction

Soil is a fragile resource that requires time to recover. Without soil, agricultural production is inconceivable, and the sustenance of the well-being of people will be impossible [1]. Soil erosion is the combination of natural and anthropogenic processes foremost to changes in soil functions in the geosystem, quantitative and qualitative degradation of soil composition, properties, and regimes, and reduction of the natural and economic importance of lands [2]. Soil erosion by water is the most common type of soil erosion, affecting about 11 million km^2 in the world [3]. Among the continents, Asia ranks first in soil erosion severity [3,4]. Soil degradation also brings enormous economic damage, disrupts the ecological balance and worsens social conditions of people [5]. The scientifically grounded and rational use of lands largely depends on the correct identification and establishment of the degree or category of erosion of the soil cover and accurate accounting for their correct nomenclature and classification. Thus, one of the most important tasks of this century is to ensure water and food security through effective agricultural productivity and reduced soil erosion [6,7].

Currently, agriculture remains an essential sector of the economy of Central Asia, providing 5.2% of gross domestic product (GDP) in Kazakhstan, 7.5% in Turkmenistan, 18.5% in Uzbekistan, 20.8% in Kyrgyzstan and 23.3% in Tajikistan [8], which employs between 20–50% of the Central Asian workforce [9]. Consequently, a better understanding of the impact of climate change on soil erosion processes is also paramount to the economy of Central Asia.

It should be noted that in recent years, climatic conditions in the Central Asian countries have changed owing to the reduction of glacier areas of the Tien Shan [10–12] and Pamir-Alay [13,14] mountain systems in the south and the drying up of the Aral Sea [15,16] in the north. In this regard, the shortage of water for irrigation, degraded natural vegetation covers, erosion processes and salinization are on the rise, while the productive capacity of irrigated lands is decreasing [8]. Humanity faces an urgent problem—the preservation of existing natural landscapes, which includes, improving and multiplying its types.

Rainfall erosivity is associated with the influential kinetic energy of raindrops, which often separates soil elements and transports them along with surface runoff [17]. Rainfall erosivity is the most significant factor and offers conservation actions by models of soil erosion prediction [18]. The rainfall erosivity (R) factor is usually adopted in soil erosion calculation models, such as the Universal Soil Loss Equation (USLE) [19] and its revised version (RUSLE) [20]. Rainfall erosivity in USLE (RUSLE) is defined as the long-term average product of total rainfall energy and maximum precipitation intensity over 30 min (EI_{30}) for storm events [19,20]. Data on the pluviograph for at least 20 years is required to calculate the original rainfall erosivity [20]. However, such data (EI_{30}) is not available in many countries and regions, and the processing of this data is quite tedious and time-consuming [21,22]. This also applies to Central Asia, where precipitation data with good temporal coverage is still scarce. However, numerous studies have established a statistical regression equation between R and variable rainfall, such as annual rainfall [4,23–25]. More recently, there has been evidence of the influence of climate change on rainfall erosivity in various parts of the globe [7,17,26–28].

Climate change is one of the most significant environmental issues of the 21st century [29,30]. Climate changes that are related to soil erosion mainly include changes in temperature and precipitation [31]. Climate change may influence rainfall erosivity alteration in precipitation patterns [17,32]. The characteristics of precipitation (amount of precipitation, its intensity and spatial-temporal distribution) directly affect soil erosion [33]. Similarly, an increase in temperature indirectly affects soil erosion [31]. The addition of water vapor to the atmosphere has an impact on the nature of climate circulation, thereby altering the intensity, frequency of extreme precipitation [7]. In arid and semi-arid climates, such as in Central Asia, there will be a more significant increase in temperature and rainfall events than in many other regions of the world [34,35].

The spatial and temporal projection of future rainfall erosivity in a changing climate in Central Asia has not been studied. Thus, the objective of this paper is to predict the value of rainfall erosivity and erosivity density in the 2030s and 2070s. We used WorldClim data [36] and the climate projections from four GCMs, Beijing Climate Center, Climate System Model 1.1 (BCCCSM1.1), Institut Pierre Simon Laplace Model CM5A-LR (IPSLCM5ALR), Model for Interdisciplinary Research On Climate version 5 (MIROC5), and the Max Planck Institute for Meteorology (MPIESMLR), with RCP2.6, and RCP8.5 scenarios.

2. Materials and Methods

2.1. Study Area

Central Asia occupies a vast territory on the Asian continent and includes the Kyrgyz Republic, the Republic of Tajikistan, the Republic of Turkmenistan, the Republic of Uzbekistan and the Republic of Kazakhstan entirely [37]. These five countries cover an area of 4 million km^2 (46°45′28.13″–87°21′47.81″ E, 35°5′2.24″–52°33′30.49″ N) [38], with a combined population of 65 million people [8]. The nations comprising Central Asia were once part of the Soviet Union. In physical-geographical terms, the region is a separate natural-historical region, sharply differing from adjacent areas by its natural conditions. All local types of Central Asian climates can be divided into three types: (1) The climate of the temperate zone, (2) climates of the dry subtropical zone and (3) the mountain climates of Tien Shan, Pamir-Alai, Pamir and Kopetdag with a well-marked altitudinal zonation [37]. Average annual precipitation in Central Asia is 254 mm, the minimum is 66 mm, and the maximum is 1222 mm (Figure 1a), elevation ranges from −229 to up to 7447 m (Figure 1b). Significant differences in the height of parts of this territory—from the areas lying below the ocean level to the highest mountain peaks—make here variety of climate and landscape forms. Having a long and orographically complex territory with extensive lowlands and highest mountain elevations in the south, southeast, Central Asia is characterized by a variety of climatic conditions. The climate of Central Asia is distinguished by a high continentality, marked by a great amplitude of fluctuations in air temperature and a meagre amount of precipitation [37].

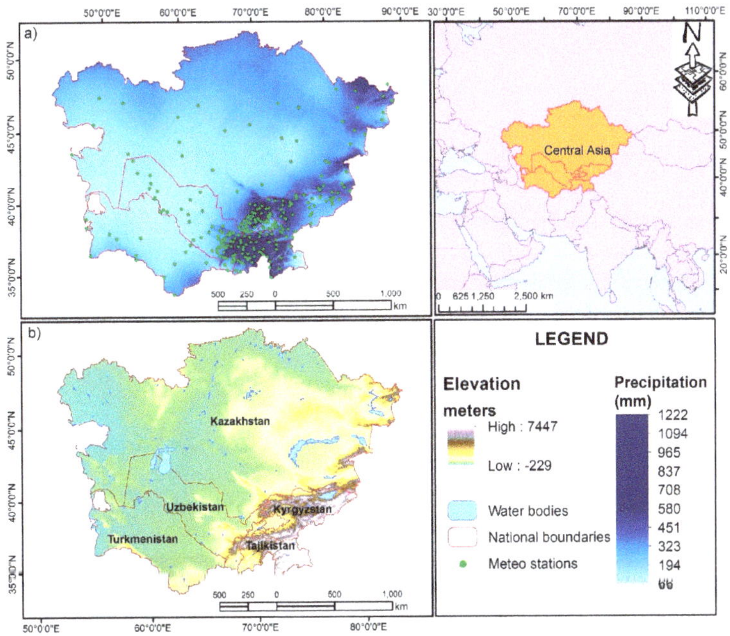

Figure 1. Study area: (**a**) Mean annual precipitation from WorldClim and (**b**) elevation from Shuttle Radar Topography Mission Digital Elevation Model (SRTM DEM, 90 m).

2.2. Climate Data

Compared to the Coupled Model Intercomparison Project Phase 3 (CMIP3), CMIP5 is a notable improvement because it uses a new set of emission scenarios called RCPs [17,39]. Projected precipitation data from GCMs BCCCSM1-1, IPSLCM5ALR, MIROC5, and MPIESMLR for the RCP2.6 and RCP8.5 greenhouse emission scenarios were used [39] (Table 1; Figure 2). The GCMs were selected because of their relative independence, good performance in precipitation simulation for Central Asia [40] and Tibetan Plateau [33]. Global precipitation with 1 km^2 horizontal resolution was obtained from the WorldClim database [36]. In assessing future changes in the erosion of rainfall and possible consequences, the predicted rainfall data for the "near" (2020–2049) and "far" future (2060–2089) have been retrieved from the Climate Change Agriculture and Food Security (CCAFS, http://www.ccafs-climate.org) portal. The data were statistically downscaled to 1 km^2 horizontal resolution using the delta method [41], based on the sum of interpolated anomalies to high-resolution monthly climate surfaces from WorldClim [36]. These anomalies were then interpolated using thin plate spline interpolation [41]. These datasets were used as input data for this study.

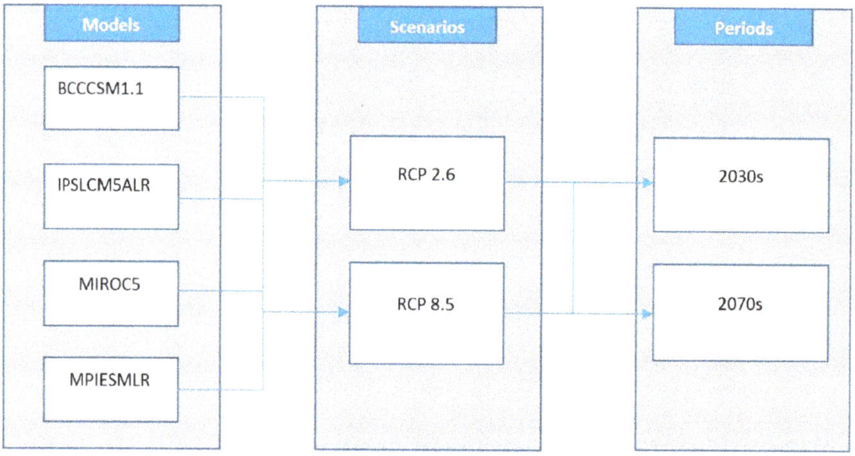

Figure 2. Climate change scenarios used in the paper.

Table 1. Global Climate Models (GCMs), from the Climate Change Agriculture and Food Security (CCAFS, http://www.ccafs-climate.org) portal.

Model	Institute	Country	Resolution
BCCCSM-1.1	Beijing Climate Center, Climate System Model 1.1	China	~2.8125° × 2.8125°
IPSLCM5ALR	Institut Pierre Simon Laplace Model, New Atmospheric Physic at Low Resolution	France	3.75° × 1.9°
MIROC-5	Model for Interdisciplinary Research On Climate	Japan	1.4° × 1.4°
MPIESMLR	Max Plank Institute for Meteorology	Germany	1.875° × ~1.9°

2.3. Estimation of Rainfall Erosivity

In this paper, the rainfall erosivity (R) factor from the RUSLE model was chosen to estimate the changes in rainfall erosivity. Rainfall erosivity was calculated using the precipitation values of gridded GCMs, comparing it with WorldClim data. [19,20] described the original method of calculating erosivity as:

$$R = \frac{1}{n}\sum_{j=1}^{n}\sum_{k=1}^{m_j}(EI_{30})k, \quad (1)$$

where R is the mean annual rainfall erosivity (MJ mm ha^{-1} h^{-1} year^{-1}), n is the number of years of data, m_j is the number of erosive events in the j year and EI_{30} is the rainfall erosivity index of a storm k. The event's rainfall erosivity index EI_{30} is defined as:

$$EI_{30} = I_{30}(\sum_{r=1}^{m} e_r v_r) \tag{2}$$

where e_r is the unit rainfall energy (MJ ha^{-1}) and v_r is the rainfall depth (mm) during a time period r. I_{30} is the maximum rainfall intensity during a 30 min period of the rainfall event (mm h^{-1}).

$$e_r = 0.29[1 - 0.072exp(-0.05i_r)] \tag{3}$$

where i_r is the rainfall intensity during the period (mm h^{-1}).

The information needed to calculate the R factor using the proposed method is usually difficult to obtain in many parts of the world. Therefore, various studies have been conducted to derive regression equations for the derivation of R factor. These simplified methods offer exceptional ease of studying the spatial and temporal variability of rainfall erosivity. Researchers [23] proposed the following equations for estimating the R factor using annual precipitation or Modified Fournier Index (*MFI*) in the absence of data on rainfall intensity for a particular site:

$$R = 0.04830 \times P^{1.61}, where\ P < 850\ \text{mm} \tag{4}$$

$$R = 587.8 - 1.219 \times P^2\ , where\ P \geq 850\ \text{mm} \tag{5}$$

where R is rainfall erosivity factor (MJ mm ha^{-1} h^{-1} year^{-1}), P is the average mean annual precipitation.

$$R = 0.7397MFI^{1.847},\ where\ MFI < 55\ \text{mm} \tag{6}$$

$$R = 95.77 - 6.081MFI + 0.4770MFI^2,\ where\ MFI \geq 55\ \text{mm} \tag{7}$$

where R is the rainfall erosivity (MJ mm ha^{-1} h^{-1} year^{-1}). *MFI* is Modified Fournier Index, given below [42,43].

$$MFI = \sum_{i=1}^{12} \frac{p_i}{P} \tag{8}$$

where P is annual precipitation (mm), and p_i is the monthly rainfall.

In this study, rainfall erosivity has been determined using the average annual precipitation, Equations (4) and (5). We used these equations because they were widely used in other similar studies [4,17]. The data used to derive R factor are gridded WorldClim data of precipitation and the GCMs.

2.4. Annual Erosivity Density Ratio

According to Kinnell [44], the erosivity density coefficient is the ratio of rainfall erosivity (R) factor to precipitation. In practice, it measures the erosivity per unit of precipitation (mm) and is expressed as MJ ha^{-1} h^{-1} (9).

$$ED = \frac{R}{P} \tag{9}$$

where ED is the erosivity density, R is the average annual rainfall erosivity and P is the average annual precipitation.

2.5. Model Evaluation Rainfall Erosivity

To evaluate the R factor of the baseline output, we made use of precipitation data from Central Asia temperature and precipitation (CATP) data (1879–2003), version 1 from the National Snow and Ice Data Center (NSIDC) [45]. This dataset contains monthly climatic data. The performance of the rainfall erosivity model was assessed by comparing the rainfall erosivity of observation data (from 269 meteorological stations) with that of the baseline data using coefficient of determination (R^2), root mean squared error (RMSE) and Nash–Sutcliff Efficiency (NSE) [46] Equations (10), (11) and (12) respectively.

$$R^2 = 1 - \left[\frac{\sum_{i=1}^{n}\left(Y_i^{mod} - Y_i^{obs}\right)^2}{\sum_{i=1}^{n}\left(Y_i^{mod} - Y_i^{obs}\right)^2 + \sum_{i=1}^{n}\left(Y_i^{mod} - Y_i^{mean}\right)^2}\right] \quad (10)$$

$$\mathrm{RMSE} = \sqrt{\frac{\sum_{i=1}^{n}\left(Y_i^{obs} - Y_i^{mod}\right)^2}{n}} \quad (11)$$

$$\mathrm{NSE} = 1 - \left[\frac{\sum_{i=1}^{n}\left(Y_i^{obs} - Y_i^{mod}\right)^2}{\sum_{i=1}^{n}\left(Y_i^{obs} - Y_i^{mean}\right)^2}\right] \quad (12)$$

where Y_i^{mod} is the baseline rainfall erosivity, Y_i^{obs} is the observed rainfall erosivity and Y_i^{mean} is the mean of observed and baseline rainfall erosivity.

3. Results

3.1. Rainfall Erosivity Under Baseline and Projected Climate

The WorldClim (baseline) and observation precipitation were statistically compared. The correlation coefficient of about 0.91 was found between the baseline and observed average monthly precipitation. The average annual rainfall erosivity for observation data ranges from 71.7–2390.3 MJ mm ha^{-1} h^{-1} $year^{-1}$ with mean 497.8 MJ mm ha^{-1} h^{-1} $year^{-1}$ and standard deviation 359 MJ mm ha^{-1} h^{-1} $year^{-1}$. While baseline data shows the range of rainfall erosivity to be 95–1838.9 MJ mm ha^{-1} h^{-1} $year^{-1}$ with mean and standard deviation 476.8 and 267.1 MJ mm ha^{-1} h^{-1} $year^{-1}$, respectively. In comparison, the baseline and observed rainfall erosivity produced 0.81, 156.7 MJ mm ha^{-1} h^{-1} $year^{-1}$ and 0.60 for R^2, RMSE and NSE, respectively. This represents a good model performance. Figure 3 illustrates the evaluation of rainfall erosivity of both observation and baseline estimates.

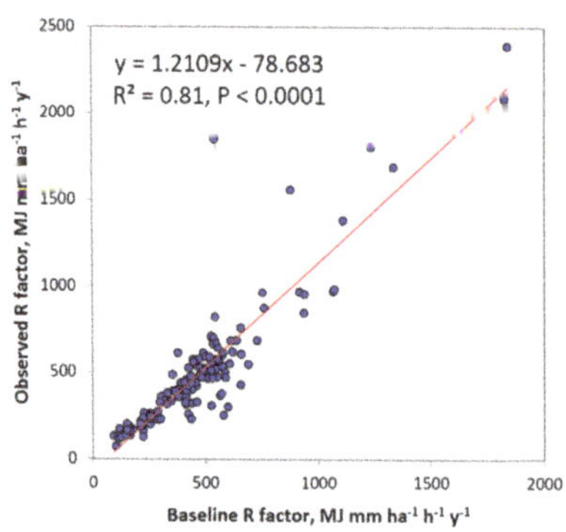

Figure 3. Scatter plot between observed and baseline annual rainfall erosivity.

The highest annual R factor was found in the southeastern part of Central Asia, with moderate values in the northern regions, but decreased westwards, where the lowest values were recorded.

On the other hand, the values gradually increased towards Tajikistan but reduced in the western parts occupied by Turkmenistan. The spatial distribution of the *R* factor constantly varied concerning annual precipitation in Central Asia. The estimated average annual rainfall erosivity for the baseline period ranges from 41 MJ mm ha^{-1} h^{-1} $year^{-1}$ to 4510 MJ mm ha^{-1} h^{-1} $year^{-1}$, in the west and southeast, respectively (Figure 4a).

The MIROC5 RCP2.6 and 8.5 show higher rainfall erosivity, perhaps due to the strongly projected spatial difference in rainfall with these scenarios. In all the GCMs and baseline precipitation, the *R* factor in Tajikistan, Kyrgyzstan, east Uzbekistan, and east Kazakhstan is higher but lower in Turkmenistan, northwest Uzbekistan, southwest and central Kazakhstan (Figure 5). Also, Figure 6 shows the relative difference between the four projected scenarios and the baseline.

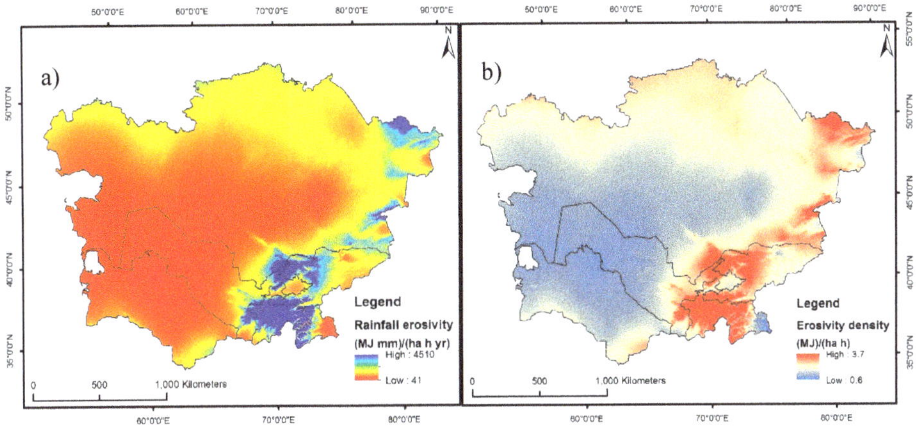

Figure 4. (**a**) Baseline rainfall erosivity and (**b**) erosivity density.

Table 2 presents the effects of rainfall on historical and projected rainfall erosivity and erosivity density in Central Asia. GCM ensembles show that rainfall erosivity increases from the baseline in all ensembles, except BCCCSM1.1-8.5 in 2070s and IPSLCM5ALR both RCPs (2030 and 2070). The average value of all scenarios shows that precipitation increased in the 2030s to 262 mm and 2070s to 268 mm, from the baseline (254 mm). Nevertheless, MIROC5 (the 2030s and 2070s, both RCP) predicted a higher increase in precipitation than other models with similar scenarios and periods.

Table 2. Changes in average rainfall erosivity and erosivity density under climate change across Central Asia.

Climate Models	Precipitation	Rainfall Erosivity (MJ mm ha^{-1} h^{-1} $year^{-1}$)	Change (%)	Erosivity Density	Change (%)
Baseline	253.57	402.07	0.0	1.38	0.0
2030s					
BCCCSM1.1-2.6	263.5	430.01	6.95	1.41	2.2
BCCCSM1.1-8.5	267.12	437.07	8.7	1.42	2.9
IPSLCM5ALR-2.6	247.31	386.65	−3.84	1.36	−1.4
IPSLCM5ALR-8.5	246.48	386.37	−3.9	1.35	−2.2
MIROC5-2.6	266.4	439.64	9.34	1.42	2.9
MIROC5-8.5	283.19	481.98	19.87	1.47	6.5
MPIESMLR-2.6	254.36	404.09	0.5	1.38	0.0
MPIESMLR-8.5	263.94	430.14	6.98	1.41	2.2
Average	261.54	424.49	5.58	1.4	1.6

Table 2. *Cont.*

Climate Models	Precipitation	Rainfall Erosivity (MJ mm ha^{-1} h^{-1} year^{-1})	Change (%)	Erosivity Density	Change (%)
2070s					
BCCCSM1.1-2.6	273.95	450.35	12.01	1.45	5.1
BCC-CSM1.1-8.5	268.61	437.77	8.88	1.43	3.6
IPSLCM5ALR-2.6	248.82	391.22	−2.7	1.36	−1.4
IPSLCM5ALR-8.5	243.9	381.36	−5.15	1.34	−2.9
MIROC5-2.6	270.33	449.88	11.89	1.43	3.6
MIROC5-8.5	294.11	508.85	26.56	1.51	9.4
MPIESMLR-2.6	278.9	469.3	16.72	1.46	5.8
MPIESMLR-8.5	267.4	435.84	8.4	1.42	2.9
Average	268.25	440.57	9.58	1.43	3.3

Precipitation, erosivity and density differ accordingly given that the GCMs exhibited consistent variations as shown in Table 2. Although average precipitation and rainfall erosivity demonstrate a steady increase in all the GCMs in combination with the baseline precipitation output, IPSLCM5ALR however shows a decline in average precipitation and erosivity in both periods.

3.2. *Rainfall Erosivity at the National Level*

In the baseline period, Kyrgyzstan had an estimated average rainfall erosivity of 869.7 MJ mm ha^{-1} h^{-1} year^{-1}. The MIROC5 and MPIESMLR scenarios (RCP2.6 and RCP8.5), respectively project an increase in the mean rainfall erosivity ranged from 27.9–50.1% and from 0.9–27%. The BCCCSM1.1 scenarios also projected a mean increase of 6.3% for all periods and a decrease (−7.8%) for the BCCCSM1.1-8.5 in the 2070s.

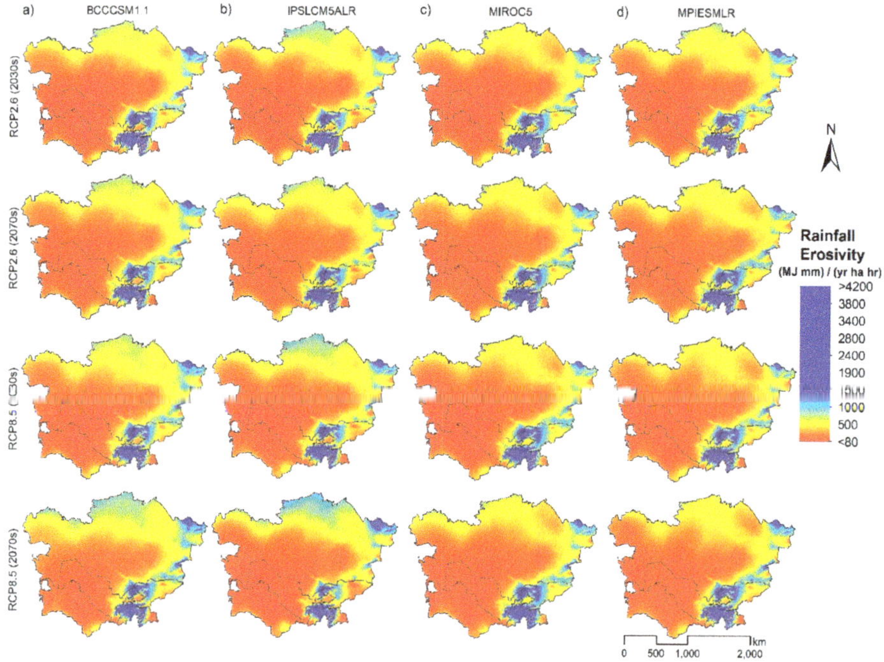

Figure 5. Rainfall erosivity projections for the period 2030s and 2070s according to RCP2.6 and 8.5 scenarios driven by the BCCCSM1.1, IPSLCM5ALR, MIROC-5, MPIESMLR GCM models.

In Kazakhstan, we calculated the mean values of rainfall erosivity of 374.3 MJ mm ha^{-1} h^{-1} year^{-1} during the baseline period. From MIROC5, BCCCSM1.1 and MPIESMLR (both scenarios), we noted an increase of 5.3% to 27.3%, 11.6% to 23.2 and 0.6% to 15.7%, respectively, in mean rainfall erosivity in this country with a slight increase in the northern part and a substantial increase in the eastern part of this country. We also found results using IPSLCM5ALR for both scenarios, with decreases ranging from −0.3% to 1.6%, except RCP8.5 (the 2070s) with an increase of 1.5%.

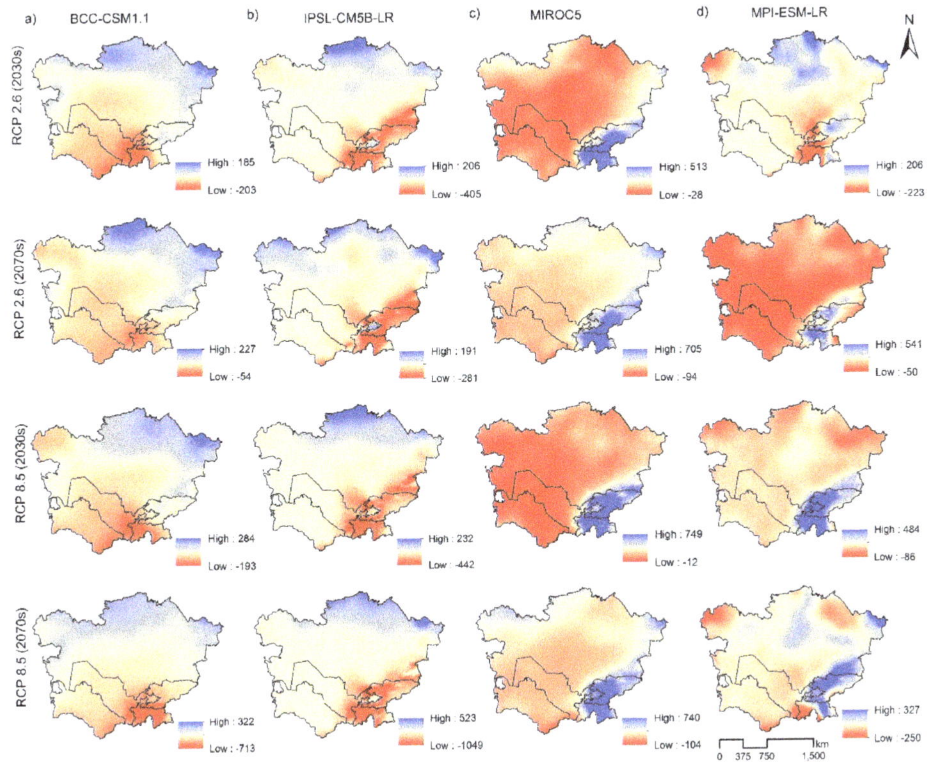

Figure 6. Absolute differences of rainfall erosivity (MJ mm ha^{-1} h^{-1} year^{-1}) between 2030, 2070 projections and baseline data.

The average rainfall erosivity in Turkmenistan for the baseline period was 188.4 MJ mm ha^{-1} h^{-1} year^{-1}. This country has the lowest rainfall erosivity among all Central Asian countries. The ensemble scenarios of IPSLCM5ALR and BCCCSM1.1 (RCP2.6 and RCP8.5) predict decrease from −10.1% to −19%, and from −7.1% to −16% respectively. However, BCCCSM1.1-2.6 (2070s) predicts about 10.5% increase in rainfall erosivity. The MPIESMLR and MIROC5 results indicate changes from 2.3% to 14.3, and from 3.7% to 19.6%, while MIROC5-2.6 (2070s) decreased (−4.7%) during the two-time slices for the two emission scenarios.

For Tajikistan, an average rainfall erosivity of 1447.7 MJ mm ha^{-1} h^{-1} year^{-1} in the baseline period was revealed. The increase was observed in the MIROC5 and MPIESMLR scenarios thereby, indicating the highest rainfall erosivity in the study area. However, there is also a decrease in the average rainfall erosivity in this country for IPSLCM5ALR (both scenarios and both periods), BCCCSM1.1-8.5 (both periods) and MPIESMLR-2.6 (the 2030s) compared to the baseline. For both scenarios, MIROC5 projected increases in erosivity from 34.3% to 56.3%. IPSLCM5ALR estimated decreases from −8.2% to −26.2%.

Uzbekistan had an average baseline rainfall erosivity of 282.1 MJ mm ha^{-1} h^{-1} year^{-1}. The MIROC5 scenarios (RCP2.6 and RCP8.5) projected an average increase ranging from 2.1–24.1%. The MPIESMLR projected a mean increase ranging from 7–15.3%, except for a decrease of −2.7% for the RCP2.6 (the 2030s). IPSLCM5ALR projected decreases in the mean annual rainfall erosivity from −7.7% to 28.2%. In general, all scenarios estimated an increase and decrease in rainfall erosivity over Uzbekistan (Table 3). Our result (<280 MJ mm ha^{-1} h^{-1} year^{-1}) is comparable with [18], which reported low average erosivity values (<250 MJ mm ha^{-1} h^{-1} year^{-1}) in Kazakhstan, Turkmenistan, and Uzbekistan.

Table 3. Rainfall erosivity in Central Asia by country. Mean baseline and estimated (MJ mm ha^{-1} h^{-1} year^{-1}) by BCCCSM1.1, IPSLCM5ALR, MIROC5 and MPIESMLR with RCP2.6 and RCP8.5 emission scenarios models. Projected change to baseline (%).

		KGZ	KZT	TJK	TKM	UZB
Baseline (1961–1990)		869.7	374.3	1447.7	188.4	282.1
RCP2.6 (2030s)	BCCCSM-1.1	903.8	420.3	1395.9	167.7	258
	Change, %	6.3	11.6	0.4	−10.1	-8.1
	IPSLCM5ALR	744.1	377.5	1290.8	164.7	240.5
	Change, %	−17.2	-0.3	−11	−11.2	−11.9
	MIROC5	1057	400.6	1697.1	196.9	295.6
	Change, %	27.9	5.3	36.9	3.7	2.1
	MPIESMLR	870.1	380.8	1388.3	191.2	272.5
	Change, %	0.9	0.6	−2.2	2.3	-2.7
RCP2.6 (2070s)	BCCCSM-1.1	910	436.2	1450.2	200.9	291.9
	Change, %	6.3	17.7	3.8	10.5	9
	IPSLCM5ALR	769	375.1	1350.5	172.9	263.4
	Change, %	−14.5	−1.5	-8.2	−7.1	−7.7
	MIROC5	1062.6	416.8	1699.9	179.9	297
	Change, %	28.8	9.8	34.3	-4.7	2.4
	MPIESMLR	1081.9	432.5	1702.5	216.9	329
	Change, %	27	15.7	24.5	14.3	15.3
RCP8.5 (2030s)	BCCCSM-1.1	909.6	430.9	1386.2	165	256.6
	Change, %	6.3	14.9	−5.7	−11.1	−10
	IPSLCM5ALR	736.4	380.2	1254.8	171	227.6
	Change, %	−17.4	−1.6	−15.4	−7.3	−17.7
	MIROC5	1216.8	435.9	1818.4	212	336.9
	Change, %	49.7	17	51.3	13.5	19.4
	MPIESMLR	1041.7	385.2	1664.4	202.8	320
	Change, %	22	4.6	20.8	8.8	14.5
RCP8.5 (2070s)	BCCCSM-1.1	794	456.8	1150.6	146.9	228.4
	Change, %	−7.8	23.2	−26.3	19	17.7
	IPSLCM5ALR	581.3	400	1082	154.2	182.8
	Change, %	33.5	1.5	−26.2	−16.7	−28.2
	MIROC5	1224.1	473.1	1800.4	215.3	340
	Change, %	50.1	27.3	56.3	19.6	24.1
	MPIESMLR	988	409.6	1484.7	195.4	294.5
	Change, %	15.7	11.1	9	7.6	7

KGZ-Kyrgyzstan, KZT-Kazakhstan, TJK-Tajikistan, TKM-Turkmenistan, UZB-Uzbekistan.

3.3. *Annual Erosivity Density*

Separately projected annual erosivity results (Figure 5) are divided by a corresponding average rainfall data to derive average erosivity density ratio. Density values of erosivity above 1 suggest that a certain amount of precipitation may lead to relatively higher rainfall erosivity [47]. The annual erosivity density for baseline period has a mean value of 1.38 MJ ha^{-1} h^{-1}, with variability ranging from 0.62–3.69 MJ ha^{-1} h^{-1} (Figure 4b). The projected variation of erosivity density is also very high as the MIROC5-8.5 has the highest mean erosivity density with 1.47 and 1.51 MJ ha^{-1} h^{-1} in the 2030s and 2070s, respectively. Followed by MPIESMLR-2.6 with 1.46 MJ ha^{-1} h^{-1} in 2070s, BCCCSM1.1-2.6 and BCCCSM1.1-8.5 with 1.45 and 1.43 in 2070s. However, IPSLCM5ALR-2.6 and IPSLCM5ALR-8.5 (both periods) have the lowest mean erosivity density with an average of 1.35 MJ ha^{-1} h^{-1} (Table 2; Figures 7 and 8).

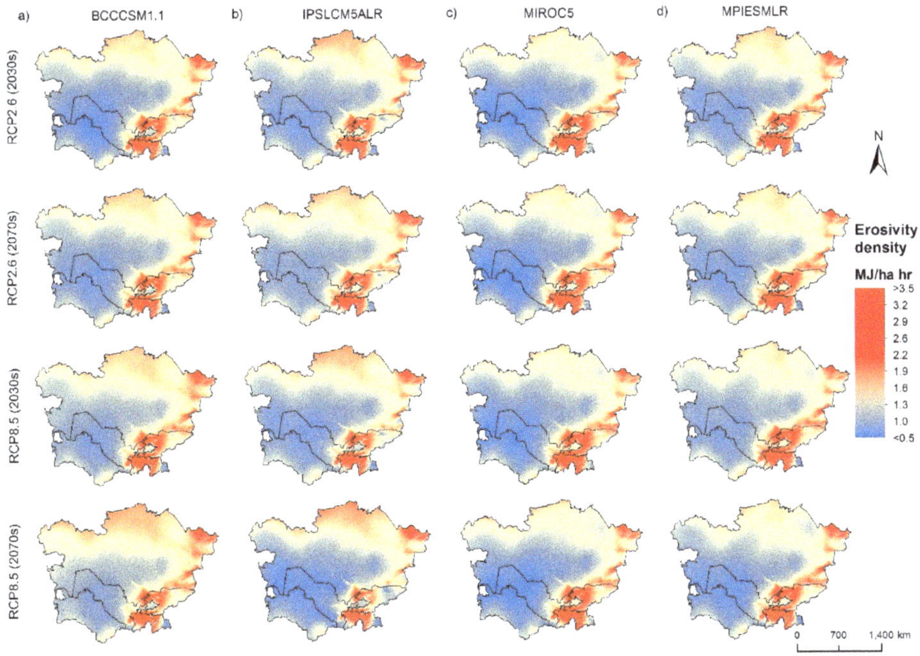

Figure 7. Erosivity density for different scenarios for the period 2030s and 2070s according to RCP2.6 and 8.5 scenarios driven by the BCCCSM1.1, IPSLCM5ALR, MIROC-5, MPIESMLR GCM models.

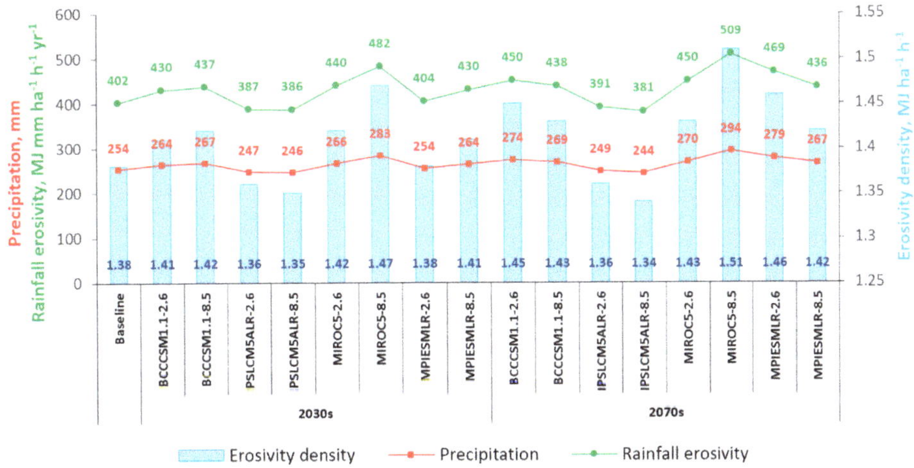

Figure 8. Annual precipitation, *R* factor and erosivity density for different scenarios and periods.

4. Discussion

The influence of climate change on rainfall erosivity is expressed by variations in total precipitation, as shown in the result section. The Tien Shan, Pamir-Alay and Pamir mountains experience more torrential rainfall compared to the surrounding low-lying deserts. Changes in precipitation mainly depend on changes in the water content in the atmosphere, which is transferred from the oceans to the earth through large-scale atmospheric circulation [40]. Atmospheric circulation over Central Asia is characterized by the predominance of the west-east transfer of air masses when the main moisture that gives precipitation comes from the North Atlantic Ocean [37,48–50]. Most of the low-latitude region (40° N) is marked by low-pressure anomalies [50]. As the air masses move from the Atlantic Ocean, they lose moisture to become dry air mass as they approach the territory—causing little or no precipitation in summer [37]. El Niño Southern Oscillation (ENSO) has affected precipitation changes over the arid regions of Central Asia by the southwestward flow of water vapor coming from the Arabian Sea and tropical Africa [48,49,51]. ENSO-induced precipitation is related to large-scale atmospheric circulation changes caused by sea surface temperature (SST) [52]. Previous studies have shown that changes in SST have a significant impact on the transport of water vapor from the oceans to land [40,53]. The main feature in the distribution of precipitation in Central Asia is their small annual amount of the lower part of the territory, resulting in vast deserts. At the same time, on the shores of the Caspian Sea, and especially Balkhash Lake, precipitation is generally low. Only in the mountains—on the outer windward slopes, where the air masses experience a forced rise, resulting to cooling, reaching a state of saturation—does the orographic increase in precipitation occur 3–5 times or more compared to the surrounding deserts [37]. This fact explains the spatial distribution of precipitation in Central Asia and, in turn, may clarify why some parts have higher erosion than other parts. This results in precipitation variability that consequently influences erosion.

As stated by [17], erosion will be affected by changes in precipitation patterns and quantities due to climate change. Studies in the Eurasian continent predicted a significant increase or decrease in erosivity for the future climate. For example, [26] found that 81% of the territory in Europe is projected to have an increase in rainfall erosivity and 19% rainfall erosivity projected to decrease by 2050 (HadGEM2, RCP4.5 scenario). Likewise, however, our study predicts some spatial variability in erosivity for Central Asia concerning the anthropogenic influence on the amount of precipitation based on different GCMs (Tables 2 and 3, Figures 5, 6 and 8).

Also, the dynamic influence of climate change on soil erosion is another essential factor that is uncertain; however, it may depend on the interacting impacts of the associated factors. Nonetheless, future soil erosion rates are expected to increase due to increased precipitation and rainfall erosivity. Moreover, this has been confirmed in other prediction studies (e.g., [17,27,28,54]), an increase in precipitation and intensity will significantly impact soil erosion rates. The highest percentage of rainfall erosivity occurs in medium and high regions of Tajikistan, Kyrgyzstan, Eastern Uzbekistan, and East Kazakhstan. Besides, this suggests that there will be more occasions for soil losses at medium and high altitudes than has ever been experienced in the past. Consequently, high soil erosion may lead to high sedimentation in rivers, lakes, and reservoirs, and these are critical for flooding and water pollution [4]. Small percent variations are usually expected in developed areas, which generally have gentle slopes and less hilly areas. The variability of our results shows the disagreements in scenarios, periods and climate models, but may also show persistent uncertainty in our models.

Preservation of fertile soils by agricultural lands, pastures, and forests is the primary condition for the sustainable development of humanity. The possible increase in rainfall erosivity in Tajikistan and Kyrgyzstan may affect a significant part of agricultural production in Central Asia due to increased soil loss and reduced soil fertility and water availability. On the other hand, a reduction in the rainfall erosivity over Turkmenistan and western Uzbekistan can strengthen the trend of agricultural development in these areas. However, climate change can significantly affect land cover, which can balance or reinforce some erosion trends. To predict future soil erosion trends, these feedbacks between precipitation and land cover should be evaluated.

5. Conclusions

In this research, we estimated the potential influence of climate change on rainfall erosivity and erosivity density in Central Asia using baseline data (1960–1990) and projected precipitation data (2020–2049 and 2060–2089). The projected precipitation was taken from the downscaled data of four GCM, BCCCSM1.1, IPSLCM5ALR, MIROC5, and MPIESMLR, based on two scenarios, RCP2.6, and RCP8.5. The mean erosion of rainfall in Central Asia was calculated and compared with climate scenario predictions. The mean rainfall erosivity within the baseline period was 497.8 MJ mm ha^{-1} h^{-1} $year^{-1}$, as compared to 476.8 MJ mm ha^{-1} h^{-1} $year^{-1}$ from the observed rainfall data for 269 meteorological stations points. The Tajikistan and Kyrgyzstan are predicted to be the most affected countries regarding rainfall erosivity. Increasing trends in annual rainfall erosivity from baseline climate up to the GCMs and the climate scenarios experienced variations in rainfall erosivity. There is a positive change in the average annual rainfall erosivity of 5.6% and 9.6% in the 2030s and 2070s respectively compared to the baseline (1960–1990).

The BCCCSM1.1 scenarios projected both increases and decreases in mean rainfall erosivity in Kyrgyzstan (−7.8% to 6.3%), Tajikistan (−26.3% to 3.8%), Turkmenistan (−19% to 10.5%), Uzbekistan (−17.7% to 9%) and increases in Kazakhstan (11.6% to 23.2%). The IPSLCM5ALR scenarios project decreases in mean rainfall erosivity in Kyrgyzstan (−14.5% to −35.5%), Tajikistan (−8.2% to −26.2%), Turkmenistan (−7.1% to −16.7%), Uzbekistan (−7.7% to −28.2%) and both increases and decreases in Kazakhstan (−1.6% to 1.5%). The MIROC5 scenarios project increases in Kyrgyzstan (27.9% to 50.1%), Kazakhstan (5.3% to 27.3%), Tajikistan (34.3% to 56.3%), Uzbekistan (2.1% to 24.1%), and both increases and decrease in Turkmenistan (−4.7% to 19.6%). The MPIESMLR scenarios project in mean rainfall erosivity increases in Kyrgyzstan (0.9% to 27%), Kazakhstan (0.6% to 15.7%), Turkmenistan (2.3% to 14.3%) and both increases in decreases in Tajikistan (−2.2% to 24.4%), Uzbekistan (−2.7% to 15.3%). The average values of erosion variations presented in this study are average changes in countries, while within these countries we found both increases and decreases, which emphasize some spatial variability of rainfall erosivity and soil erosion in Central Asia.

The aggregate average annual precipitation and erosion activity for all climate models for all scenarios show steady growth compared with the baseline climate, only IPSLCM5ALR (RCP2.6 and 8.5) shows a decrease in the average erosivity for the 2030 and 2070 scenarios. Higher amounts of rainfall were the main factor for the spatiotemporal variability in rainfall erosivity. Public policies aimed at preserving soil and water resources should be encouraged and applied at the national land survey level. Further study is required to consider other essential influences that intensify the erosivity; particularly the future land cover changes.

Author Contributions: E.D. carried out conceptualization, data collection, model simulation; X.C. supervised this study, E.D. and A.C.A. developed the methodology; E.D., F.U.O., R.O., A.C.A., G.I. and G.O. wrote the manuscript. All authors read, edited the final version of this manuscript.

Funding: This study supported by the Strategic Priority Research Program of the Chinese Academy of Sciences, Pan-Third Pole Environment Study for a Green Silk Road (Grant No. XDA20060303) and Funds for International Cooperation and Exchange of the National Natural Sciences of China (Grant No. 41761144079).

Acknowledgments: The authors would express sincere gratitude to the WorldClim data portal, Climate Change Agriculture and Food Security (CCAFS) and the National Snow and Ice Data Center (NSIDC) for providing climate data.

Conflicts of Interest: The authors declare no conflict of interest.

References

1. Mamytov, A.M.; Roychenko, G.I. *Soil Zoning of Kyrgyzstan*; Izd-vo AN Kirg. Frunze, Kyrgyzstan, 1961. (In Russian)
2. Khitrov, N.B.; Ivanov, A.L.; Zavalin, A.A.; Kuznetsov, M.S. Problems of Degradation, Protection and Ways of Recovery Productivity of Agricultural Land. *Vestnik Orel GAU* **2007**, *6*, 29–32. (In Russian)

3. Oldeman, L. The global extent of soil degradation1. In *Szabolcs I, Eds1 Soil Resilience and Sustainable Land Use1*; Greenland, D.J., Ed.; CAB International: Wallingford, UK, 1994; pp. 99–1181.
4. Yang, D.; Kanae, S.; Oki, T.; Koike, T.; Musiake, K. Global potential soil erosion with reference to land use and climate changes. *Hydrol. Process.* **2003**, *17*, 2913–2928. [CrossRef]
5. Dobrovolsky, G.V. *Degradation and Protection of Soils*; Dorovolsky, G.V., Ed.; Moscow State University Publishing House: Moscow, Russia, 2002; p. 654.
6. Oliveira, P.T.S.; Nearing, M.A.; Wendland, E. Orders of magnitude increase in soil erosion associated with land use change from native to cultivated vegetation in a Brazilian savannah environment. *Earth Surf. Process. Landf.* **2015**, *40*, 1524–1532. [CrossRef]
7. Almagro, A.; Oliveira, P.T.S.; Nearing, M.A.; Hagemann, S. Projected climate change impacts in rainfall erosivity over Brazil. *Sci. Rep.* **2017**, *7*, 8130. [CrossRef] [PubMed]
8. Hamidov, A.; Helming, K.; Balla, D. Impact of agricultural land use in Central Asia: A review. *Agron. Sustain. Dev.* **2016**, *36*, 6. [CrossRef]
9. Qushimov, B.; Ganiev, I.; Rustamova, I.; Haitov, B.; Islam, K. Land degradation by agricultural activities in Central Asia. In *Climate Change and Terrestrial Carbon Sequestration in Central Asia*; Taylor & Francis: London, UK, 2007; pp. 137–147.
10. Duishonakunov, M.; Imbery, S.; Narama, C.; Mohanty, A.; King, L. Recent glacier changes and their impact on water resources in Chon and Kichi Naryn Catchments, Kyrgyz Republic. *Water Sci. Technol. Water Supply* **2014**, *14*, 444–452. [CrossRef]
11. Aizen, V.B.; Kuzmichenok, V.A.; Surazakov, A.B.; Aizen, E.M. Glacier changes in the Tien Shan as determined from topographic and remotely sensed data. *Glob. Planet. Chang.* **2007**, *56*, 328–340. [CrossRef]
12. Kenzhebaev, R.; Barandun, M.; Kronenberg, M.; Chen, Y.; Usubaliev, R.; Hoelzle, M. Mass balance observations and reconstruction for Batysh Sook Glacier, Tien Shan, from 2004 to 2016. *Cold Reg. Sci. Technol.* **2017**, *135*, 76–89. [CrossRef]
13. Chevallier, P.; Pouyaud, B.; Mojaïsky, M.; Bolgov, M.; Olsson, O.; Bauer, M.; Froebrich, J. River flow regime and snow cover of the Pamir Alay (Central Asia) in a changing climate. *Hydrol. Sci. J.* **2014**, *59*, 1491–1506. [CrossRef]
14. Hagg, W.; Braun, L.N.; Kuhn, M.; Nesgaard, T.I. Modelling of hydrological response to climate change in glacierized Central Asian catchments. *J. Hydrol.* **2007**, *332*, 40–53. [CrossRef]
15. Lioubimtseva, E.; Henebry, G.M. Climate and environmental change in arid Central Asia: Impacts, vulnerability, and adaptations. *J. Arid Environ.* **2009**, *73*, 963–977. [CrossRef]
16. Issanova, G.; Jilili, R.; Abuduwaili, J.; Kaldybayev, A.; Saparov, G.; Yongxiao, G. Water availability and state of water resources within water-economic basins in Kazakhstan. *Paddy Water Environ.* **2018**, *16*, 183–191. [CrossRef]
17. Amanambu, A.C.; Li, L.; Egbinola, C.N.; Obarein, O.A.; Mupenzi, C.; Chen, D. Spatio-temporal variation in rainfall-runoff erosivity due to climate change in the Lower Niger Basin, West Africa. *Catena* **2019**, *172*, 324–334. [CrossRef]
18. Panagos, P.; Borrelli, P.; Meusburger, K.; Yu, B.; Klik, A.; Jae Lim, K.; Yang, J.E.; Ni, J.; Miao, C.; Chattopadhyay, N.; et al. Global rainfall erosivity assessment based on high-temporal resolution rainfall records. *Sci. Rep.* **2017**, *7*, 4175. [CrossRef] [PubMed]
19. Wischmeier, W.H.; Smith, D.D. Predicting rainfall erosion losses-a guide to conservation planning. In *Predicting Rainfall Erosion Losses—A Guide to Conservation Planning*; USDA, Science and Education Administration: Hyattsville, MD, USA, 1978.
20. Renard, K.G.; Foster, G.R.; Weesies, G.; McCool, D.; Yoder, D. *Predicting Soil Erosion by Water: A Guide to Conservation Planning with the Revised Universal Soil Loss Equation (RUSLE)*; United States Department of Agriculture: Washington, DC, USA, 1997; Volume 703.
21. Lai, C.; Chen, X.; Wang, Z.; Wu, X.; Zhao, S.; Wu, X.; Bai, W. Spatio-temporal variation in rainfall erosivity during 1960–2012 in the Pearl River Basin, China. *Catena* **2016**, *137*, 382–391. [CrossRef]
22. Lee, J.-H.; Heo, J.-H. Evaluation of estimation methods for rainfall erosivity based on annual precipitation in Korea. *J. Hydrol.* **2011**, *409*, 30–48. [CrossRef]
23. Renard, K.G.; Freimund, J.R. Using monthly precipitation data to estimate the R-factor in the revised USLE. *J. Hydrol.* **1994**, *157*, 287–306. [CrossRef]

24. Angulo-Martínez, M.; Beguería, S. Estimating rainfall erosivity from daily precipitation records: A comparison among methods using data from the Ebro Basin (NE Spain). *J. Hydrol.* **2009**, *379*, 111–121. [CrossRef]
25. Naipal, V.; Reick, C.H.; Pongratz, J.; Van Oost, K. Improving the global applicability of the RUSLE model-adjustment of the topographical and rainfall erosivity factors. *Geosci. Model Dev.* **2015**, *8*, 2893–2913. [CrossRef]
26. Panagos, P.; Ballabio, C.; Meusburger, K.; Spinoni, J.; Alewell, C.; Borrelli, P. Towards estimates of future rainfall erosivity in Europe based on REDES and WorldClim datasets. *J. Hydrol.* **2017**, *548*, 251–262. [CrossRef]
27. Plangoen, P.; Babel, M.S.; Clemente, R.S.; Shrestha, S.; Tripathi, N.K. Simulating the impact of future land use and climate change on soil erosion and deposition in the Mae Nam Nan sub-catchment, Thailand. *Sustainability* **2013**, *5*, 3244–3274. [CrossRef]
28. Gupta, S.; Kumar, S. Simulating climate change impact on soil erosion using RUSLE model– A case study in a watershed of mid-Himalayan landscape. *J. Earth Syst. Sci.* **2017**, *126*, 43. [CrossRef]
29. Campbell, J.L.; Driscoll, C.T.; Pourmokhtarian, A.; Hayhoe, K. Streamflow responses to past and projected future changes in climate at the Hubbard Brook Experimental Forest, New Hampshire, United States. *Water Resour. Res.* **2011**, *47*. [CrossRef]
30. Carter, J.G.; Cavan, G.; Connelly, A.; Guy, S.; Handley, J.; Kazmierczak, A. Climate change and the city: Building capacity for urban adaptation. *Prog. Plan.* **2015**, *95*, 1–66. [CrossRef]
31. Li, Z.; Fang, H. Impacts of climate change on water erosion: A review. *Earth-Sci. Rev.* **2016**, *163*, 94–117. [CrossRef]
32. Mondal, A.; Khare, D.; Kundu, S. Change in rainfall erosivity in the past and future due to climate change in the central part of India. *Int. Soil Water Conserv. Res.* **2016**, *4*, 186–194. [CrossRef]
33. Teng, H.; Liang, Z.; Chen, S.; Liu, Y.; Rossel, R.A.V.; Chappell, A.; Yu, W.; Shi, Z. Current and future assessments of soil erosion by water on the Tibetan Plateau based on RUSLE and CMIP5 climate models. *Sci. Total Environ.* **2018**, *635*, 673–686. [CrossRef]
34. Immerzeel, W.; Pellicciotti, F.; Bierkens, M. Rising river flows throughout the twenty-first century in two Himalayan glacierized watersheds. *Nat. Geosci.* **2013**, *6*, 742–745. [CrossRef]
35. Unger-Shayesteh, K.; Vorogushyn, S.; Merz, B.; Frede, H.-G. Introduction to "water in Central Asia—Perspectives under global change". *Glob. Planet. Chang.* **2013**, *100*, 1–152. [CrossRef]
36. Hijmans, R.J.; Cameron, S.E.; Parra, J.L.; Jones, P.G.; Jarvis, A. Very high resolution interpolated climate surfaces for global land areas. *Int. J. Climatol.* **2005**, *25*, 1965–1978. [CrossRef]
37. Alamanov, S.; Lelevkin, V.; Podrezov, O.; Podrezov, A. *Climate Changes and Water Problems in Central Asia*; UNEP and WWF: Moscow, Russia, 2006. (In Russian)
38. Chen, X.; Zhou, Q. *Ecological and Environmental Remote Sensing in Arid Zone–A Case Study of Central Asia*; Science Press: Beijing, China, 2015.
39. Taylor, K.E.; Stouffer, R.J.; Meehl, G.A. An overview of CMIP5 and the experiment design. *Bull. Am. Meteorol. Soc.* **2012**, *93*, 485–498. [CrossRef]
40. Luo, M.; Liu, T.; Meng, F.; Duan, Y.; Bao, A.; Frankl, A.; De Maeyer, P. Spatiotemporal characteristics of future changes in precipitation and temperature in Central Asia. *Int. J. Climatol.* **2019**, *39*, 1571–1588. [CrossRef]
41. Ramirez-Villegas, J.; Jarvis, A. *Downscaling Global Circulation Model Outputs: The Delta Method Decision and Policy Analysis Working Paper No. 1*; International Center for Tropical Agriculture (CIAT): Cali, Colombia, 2010.
42. Arnoldus, H. *Methodology Used to Determine the Maximum Potential Average Annual Soil Loss Due to Sheet and Rill Erosion in Morocco*; FAO Soils Bulletins (FAO): Rome, Italy, 1977.
43. Arnoldus, H. An approximation of the rainfall factor in the Universal Soil Loss Equation. In *An Approximation of the Rainfall Factor in the Universal Soil Loss Equation*; John Wiley and Sons Ltd.: Chichester, UK, 1980; pp. 127–132.
44. Kinnell, P. Event soil loss, runoff and the Universal Soil Loss Equation family of models: A review. *J. Hydrol.* **2010**, *385*, 384–397. [CrossRef]
45. Williams, M.; Konovalov, V. *Central Asia Temperature and Precipitation Data, 1879–2003*; USA National Snow and Ice Data Center: Boulder, CO, USA, 2008.
46. Nash, J.E.; Sutcliffe, J.V. River flow forecasting through conceptual models part I—A discussion of principles. *J. Hydrol.* **1970**, *10*, 282–290. [CrossRef]
47. Panagos, P.; Ballabio, C.; Borrelli, P.; Meusburger, K. Spatio-temporal analysis of rainfall erosivity and erosivity density in Greece. *Catena* **2016**, *137*, 161–172. [CrossRef]

48. Chen, X.; Wang, S.; Hu, Z.; Zhou, Q.; Hu, Q. Spatiotemporal characteristics of seasonal precipitation and their relationships with ENSO in Central Asia during 1901–2013. *J. Geogr. Sci.* **2018**, *28*, 1341–1368. [CrossRef]
49. Hu, Z.; Zhou, Q.; Chen, X.; Qian, C.; Wang, S.; Li, J. Variations and changes of annual precipitation in Central Asia over the last century. *Int. J. Climatol.* **2017**, *37*, 157–170. [CrossRef]
50. Hu, Z.; Chen, X.; Chen, D.; Li, J.; Wang, S.; Zhou, Q.; Yin, G.; Guo, M. "Dry gets drier, wet gets wetter": A case study over the arid regions of central Asia. *Int. J. Climatol.* **2018**, *39*, 1072–1091. [CrossRef]
51. Mariotti, A. How ENSO impacts precipitation in southwest central Asia. *Geophys. Res. Lett.* **2007**, *34*. [CrossRef]
52. Dai, A.; Wigley, T. Global patterns of ENSO-induced precipitation. *Geophys. Res. Lett.* **2000**, *27*, 1283–1286. [CrossRef]
53. Liu, S.; Duan, A. Impacts of the global sea surface temperature anomaly on the evolution of circulation and precipitation in East Asia on a quasi-quadrennial cycle. *Clim. Dyn.* **2017**, *51*, 4077–4094. [CrossRef]
54. Litschert, S.E.; Theobald, D.M.; Brown, T.C. Effects of climate change and wildfire on soil loss in the Southern Rockies Ecoregion. *Catena* **2014**, *118*, 206–219. [CrossRef]

MDPI
St. Alban-Anlage 66
4052 Basel
Switzerland
Tel. +41 61 683 77 34
Fax +41 61 302 89 18
www.mdpi.com

Water Editorial Office
E-mail: water@mdpi.com
www.mdpi.com/journal/water